TEACHING
SCIENCE
THROUGH
TRADE BOOKS

TEACHING
SCIENCE
THROUGH
TRADE BOOKS

By Christine Anne Royce, Emily Morgan, and Karen Ansberry

National Science Teachers Association

Arlington, Virginia

National Science Teachers Association

Claire Reinburg, Director
Jennifer Horak, Managing Editor
Andrew Cooke, Senior Editor
Wendy Rubin, Associate Editor
Agnes Bannigan, Associate Editor
Amy America, Book Acquisitions
    Coordinator

SCIENCE AND CHILDREN
Linda Froschauer, Editor
Valynda Mayes, Managing Editor
Stephanie Andersen, Associate
    Editor

ART AND DESIGN
Will Thomas Jr., Director
Linda Olliver, Cover and Interior
    Design and Illustration

PRINTING AND PRODUCTION
Catherine Lorrain, Director

NATIONAL SCIENCE TEACHERS ASSOCIATION
Francis Q. Eberle, PhD, Executive Director
David Beacom, Publisher

1840 Wilson Blvd., Arlington, VA 22201
www.nsta.org/store
For customer service inquiries, please call 800-277-5300.

Library of Congress Cataloging-in-Publication Data
Royce, Christine Anne, 1966-
  Teaching science through trade books / by Christine Anne Royce, Karen Ansberry, and Emily Morgan.
      p. cm.
  Includes bibliographical references and index.
  ISBN 978-1-936959-13-6 (print : alk. paper) -- ISBN 978-1-936959-86-0 (e-book : alk. paper) (print) 1.
Science--Study and teaching (Elementary) 2. Children's literature in science education. I. Ansberry, Karen
Rohrich, 1966- II. Morgan, Emily R. (Emily Rachel), 1973- III. Title.
  LB1585.R59 2012
  372.35'044--dc23
                            2012002170

Contents

Engineering, Technology, and Applications of Science

Physical Science

Life Science

Earth and Space Science

Acknowledgments

This book is the compilation of more than 10 years of work and experience by the authors in the use of children's literature to teach science. "Teaching Through Trade Books" is a monthly column that appears in *Science and Children*. We would like to acknowledge the National Science Teachers Association for their continued dedication to this approach in the teaching of elementary science. Specifically, we would like to thank the following individuals for their support, encouragement, careful editing, and critical eye in the process of writing these activities throughout the years:

- Chris Ohana, field editor for *Science and Children* from 2003 to 2009
- Linda Froschauer, current field editor for *Science and Children*
- Monica Zerry, former managing editor for *Science and Children*
- Valynda Mayes, current managing editor for *Science and Children*
- Stephanie Anderson, former associate editor for *Science and Children*

We appreciate the care, focus, and attention to detail given to this compendium project by NSTA Press managing editor Jennifer Horak.

Christine would like to acknowledge the following individuals for their contributions to her professional endeavors:

- Her mother, Ann Marie Robinson, who instilled in her three daughters the importance of education as well as reading at an early age
- Dr. David Wiley, professor, colleague, and friend who first introduced her to presenting at conferences and the professional opportunities that exist in writing
- Dr. James Johnson, dean of the College of Education and Human Services at Shippensburg University, who consistently encourages her to push new limits and supports her in professional endeavors
- Dr. Lynn Baynum, who shares every new book she gets and doesn't think it's odd to have nearly 600 children's books
- All of her friends and colleagues who are morally supportive and willing to listen to science stories

Emily and Karen would like to acknowledge the following people:

- Katie Davis, second-grade teacher at Western Row Elementary, for cowriting the "Mysteries of the Past" lessons and trying out the activities in her classroom
- Colleen Phillips-Birdsong, reading specialist and former second-grade teacher at Forest Hills Local Schools, for cowriting the "Cloud Watchers" lessons and trying out the activities in her classroom
- Claire Reinburg, director of NSTA Press, for giving them so many opportunities so share their ideas with teachers across the country
- Sue Livingston, retired teacher and curriculum specialist, who taught them that every teacher is a reading teacher
- Dr. Robert Yearout, for giving them their first opportunity to share Picture-Perfect Science with teachers and for years of encouragement and support
- Teachers Connie Farris, Kim Florek, Kathy Theiss, and Donna Hoover for the seed riddle activity that appears in Chapter 24, "Secrets of Seeds."

About the Authors

Christine Anne Royce is an associate professor of education at Shippensburg University, in Pennsylvania, where she also serves as chairperson for the teacher education department. She predominantly teaches elementary science methods classes to undergraduate students and research design at the graduate level. Prior to moving to the college level, she also taught elementary, middle, and high school students. She has a doctorate in science education from Temple University, where her dissertation focused on the use of children's literature in place of or in conjunction with textbooks. She holds a master of  science in administration and supervision from the University of Scranton, a master of arts in curriculum and instruction from Delaware State University, and a bachelor of science in elementary education from Cabrini College. She enjoys reading, photography, and genealogy. Christine currently lives near Shippensburg, Pennsylvania, with her three cats.

Emily Morgan is a consultant for Picture-Perfect Science, LLC, where she facilitates elementary science workshops for teachers nationwide. She feels that tapping into students' fascination with science will give them the motivation to read about it. Emily has a bachelor's degree in education from Wright State University and a master's in education from the University of Dayton. She taught seventh-grade science at Northridge Local Schools in Dayton, Ohio, and second- through fourth-grade science lab at Mason City Schools in Mason, Ohio.  She has served as a science consultant for the Hamilton County Educational Service Center in Cincinnati, Ohio, and as the science leader for the High AIMS Consortium. Emily is coauthor of the Picture-Perfect Science series published by NSTA Press. She lives in West Chester, Ohio with her husband, son, and an assortment of animals.

Karen Ansberry is the elementary science curriculum leader for Mason City Schools, in Mason, Ohio. As a former classroom teacher, she understands that teachers are crunched for time and need high-interest, ready-to-use lessons that integrate literature, reading strategies, and science. After graduating from Xavier University in Cincinnati, Ohio, with a bachelor's degree in biology, she completed an internship in the Cincinnati Zoo's education department.  This experience inspired her to change her career focus from wildlife biology to elementary education, and after earning a master's in teaching from Miami University in Oxford, Ohio, she began teaching fifth- and sixth-grade science at Mason City Schools. Karen is coauthor of the Picture-Perfect Science series from NSTA Press. She lives in Lebanon, Ohio, with her husband, two sons, two daughters, and too many pets.

Christine, Emily, and Karen are coauthors of the "Teaching Through Trade Books" column from NSTA's elementary school journal, *Science and Children*.

National Science Education Standards: Content Standards K–4

	Content Standard A: Science as Inquiry	Content Standard B: Physical Science	Content Standard C: Life Science	Content Standard D: Earth and Space Science	Content Standard E: Science and Technology	Content Standard F: Science in Personal and Social Perspectives	Content Standard G: History and Nature of Science
Thought-Provoking Questions	♦						
A Closer Look	♦	♦					
Science Measures Up	♦				♦		
Going Wild With Graphs	♦						
Wild About Data	♦						
Taking Note of Natural Resources						♦	
Words to the Wild							♦
Into the Woods	♦						
Discover Reading							♦
How It's Made					♦	♦	♦
It's About Time				♦			♦
If You Build It …		♦			♦		

	Content Standard A: Science as Inquiry	Content Standard B: Physical Science	Content Standard C: Life Science	Content Standard D: Earth and Space Science	Content Standard E: Science and Technology	Content Standard F: Science in Personal and Social Perspectives	Content Standard G: History and Nature of Science
Imaginative Inventions					♦	♦	
Let's Try It Out in the Air	♦	♦				♦	
A Balancing Act	♦	♦					
Gravity	♦	♦					
Roller Coasters!		♦			♦		
Secrets of Flight		♦					♦
Flick a Switch		♦				♦	
The Wonder of Water	♦	♦					
Kitchen Chemistry	♦	♦					
Secrets of Seeds	♦		♦				
Pumpkins!	♦						
Flower Power			♦				
Crazy for Loco Beans	♦		♦				
Seeing and Sorting Seashells	♦		♦				
Unusual Creatures	♦		♦				
Can You See Me Now?			♦				

	Content Standard A: Science as Inquiry	Content Standard B: Physical Science	Content Standard C: Life Science	Content Standard D: Earth and Space Science	Content Standard E: Science and Technology	Content Standard F: Science in Personal and Social Perspectives	Content Standard G: History and Nature of Science
Survival Skills			♦				
Antarctic Adaptations			♦				
The Mystery of Migration			♦				
Whoo Eats What?			♦				
A Habitat Is a Home			♦				
Exploring Your Environment			♦				
You Are What You Eat!			♦			♦	
Moving My Body			♦				
Science From the Heart	♦					♦	
Disease Detectives						♦	
Rock Solid Science				♦			
Rocking Around the Rock Cycle				♦			
The Dirt on Soil				♦			
Fascinating Fossil Finds	♦			♦			

	Content Standard A: Science as Inquiry	Content Standard B: Physical Science	Content Standard C: Life Science	Content Standard D: Earth and Space Science	Content Standard E: Science and Technology	Content Standard F: Science in Personal and Social Perspectives	Content Standard G: History and Nature of Science
Mysteries of the Past	♦			♦			
Earthquakes!				♦			
Delving Into Disasters	♦						
Cloud Watchers				♦			
Weather Watchers	♦			♦			
Sunrise, Sunset	♦			♦			
Moon Phases and Models	♦			♦			
Seeing Stars				♦			

National Science Education Standards: Content Standards 5–8

	Content Standard A: Science as Inquiry	Content Standard B: Physical Science	Content Standard C: Life Science	Content Standard D: Earth and Space Science	Content Standard E: Science and Technology	Content Standard F: Science in Personal and Social Perspectives	Content Standard G: History and Nature of Science
Thought-Provoking Questions	♦						
A Closer Look	♦						
Science Measures Up	♦				♦		
Going Wild With Graphs	♦						
Wild About Data	♦						
Taking Note of Natural Resources						♦	
Words to the Wild							♦
Into the Woods							♦
Discover Reading							♦
How It's Made					♦	♦	♦
It's About Time				♦	♦		♦
If You Build It …						♦	

	Content Standard A: Science as Inquiry	Content Standard B: Physical Science	Content Standard C: Life Science	Content Standard D: Earth and Space Science	Content Standard E: Science and Technology	Content Standard F: Science in Personal and Social Perspectives	Content Standard G: History and Nature of Science
Imaginative Inventions					♦	♦	
Let's Try It Out in the Air	♦	♦					
A Balancing Act	♦	♦					
Gravity	♦	♦					
Roller Coasters!		♦			♦		
Secrets of Flight		♦					♦
Flick a Switch		♦				♦	
The Wonder of Water	♦	♦					
Kitchen Chemistry	♦	♦					
Secrets of Seeds	♦						
Pumpkins!	♦						
Flower Power			♦				
Crazy for Loco Beans	♦		♦				
Seeing and Sorting Seashells	♦						
Unusual Creatures	♦		♦				

	Content Standard A: Science as Inquiry	Content Standard B: Physical Science	Content Standard C: Life Science	Content Standard D: Earth and Space Science	Content Standard E: Science and Technology	Content Standard F: Science in Personal and Social Perspectives	Content Standard G: History and Nature of Science
Can You See Me Now?			♦				
Survival Skills			♦				
Antarctic Adaptations			♦				
The Mystery of Migration			♦				
Whoo Eats What?			♦				
A Habitat Is a Home			♦				
Exploring Your Environment			♦				
You Are What You Eat!						♦	
Moving My Body			♦				
Science From the Heart	♦		♦			♦	
Disease Detectives			♦			♦	
Rock Solid Science				♦			
Rocking Around the Rock Cycle				♦			

	Content Standard A: Science as Inquiry	Content Standard B: Physical Science	Content Standard C: Life Science	Content Standard D: Earth and Space Science	Content Standard E: Science and Technology	Content Standard F: Science in Personal and Social Perspectives	Content Standard G: History and Nature of Science
The Dirt on Soil			♦	♦			
Fascinating Fossil Finds	♦			♦			
Mysteries of the Past	♦			♦			♦
Earthquakes!				♦	♦		
Delving Into Disasters	♦			♦			
Cloud Watchers				♦			♦
Weather Watchers	♦			♦			
Sunrise, Sunset	♦			♦			
Moon Phases and Models	♦			♦			
Seeing Stars				♦			

Alignment With *A Framework for K–12 Science Education: Practices, Crosscutting Concepts, and Core Ideas*

	Dimension #1: Scientific and Engineering Practices	Dimension #2: Crosscutting Concepts	Dimension #3: Disciplinary Core Ideas— Physical Sciences	Dimension #3: Disciplinary Core Ideas— Life Sciences	Dimension #3: Disciplinary Core Ideas— Earth and Space Sciences	Dimension #3: Disciplinary Core Ideas— Engineering, Technology, and Applications of Science
Thought-Provoking Questions	♦					
A Closer Look	♦					
Science Measures Up	♦					♦
Going Wild With Graphs	♦					
Wild About Data	♦					
Taking Note of Natural Resources					♦	
Words to the Wild	♦					
Into the Woods				♦		
Discover Reading	♦					

	Dimension #1: Scientific and Engineering Practices	Dimension #2: Crosscutting Concepts	Dimension #3: Disciplinary Core Ideas—Physical Sciences	Dimension #3: Disciplinary Core Ideas—Life Sciences	Dimension #3: Disciplinary Core Ideas—Earth and Space Sciences	Dimension #3: Disciplinary Core Ideas—Engineering, Technology, and Applications of Science
How It's Made		♦				
It's About Time		♦			♦	♦
If You Build It …		♦				♦
Imaginative Inventions		♦				♦
Let's Try It Out in the Air	♦		♦			
A Balancing Act	♦		♦			
Gravity						
Roller Coasters!			♦			♦
Secrets of Flight			♦			♦
Flick a Switch			♦			
The Wonder of Water	♦		♦			
Kitchen Chemistry	♦		♦			
Secrets of Seeds	♦					
Pumpkins!	♦					
Flower Power				♦		
Crazy for Loco Beans	♦			♦		

	Dimension #1: Scientific and Engineering Practices	Dimension #2: Crosscutting Concepts	Dimension #3: Disciplinary Core Ideas—Physical Sciences	Dimension #3: Disciplinary Core Ideas—Life Sciences	Dimension #3: Disciplinary Core Ideas—Earth and Space Sciences	Dimension #3: Disciplinary Core Ideas—Engineering, Technology, and Applications of Science
Seeing and Sorting Seashells	♦	♦				
Unusual Creatures	♦			♦		
Can You See Me Now?				♦		
Survival Skills				♦		
Antarctic Adaptations				♦		
The Mystery of Migration				♦		
Whoo Eats What?				♦		
A Habitat Is a Home				♦		
Exploring Your Environment				♦		
You Are What You Eat!				♦		
Moving My Body				♦		
Science From the Heart	♦			♦		
Disease Detectives		♦		♦		
Rock Solid Science					♦	

	Dimension #1: Scientific and Engineering Practices	Dimension #2: Crosscutting Concepts	Dimension #3: Disciplinary Core Ideas—Physical Sciences	Dimension #3: Disciplinary Core Ideas—Life Sciences	Dimension #3: Disciplinary Core Ideas—Earth and Space Sciences	Dimension #3: Disciplinary Core Ideas—Engineering, Technology, and Applications of Science
Rocking Around the Rock Cycle					♦	
The Dirt on Soil					♦	
Fascinating Fossil Finds	♦				♦	
Mysteries of the Past	♦				♦	
Earthquakes!					♦	
Delving Into Disasters	♦				♦	
Cloud Watchers					♦	
Weather Watchers	♦				♦	
Sunrise, Sunset	♦				♦	
Moon Phases and Models	♦	♦			♦	
Seeing Stars		♦			♦	

Chapter 1

Historical Use of Trade Books in the Science Classroom

"Teaching Through Trade Books" is a column that has appeared monthly in *Science and Children* since January 2003. It highlights the science and literacy connections that can be made with children's literature and provides readers with "novel" lessons that feature the newest and most engaging children's literature as well as traditional favorites. With this compendium, column authors Karen Ansberry, Emily Morgan, and Christine Royce handpicked 50 of their favorite columns, comprising 100 different activities, updated them, and organized them around the new *A Framework for K–12 Science Education: Practices, Crosscutting Concepts, and Core Ideas*. Another feature of this collection is the addition of student pages with data sheets, information tables, and more to assist teachers in their daily planning.

Each column in Teaching Science Through Trade Books follows the same basic structure:

An introduction provides a quick overview of the content or process skills that are covered with the selected trade books and accompanying activities.

Selected Children's Trade Books, which introduce the featured trade books, provide the covers for easy identification, bibliographical information for ordering the books, and a synopsis about each book.

Curricular Connections tie the activities to the National Science Education Standards and indicate where and how the selected books and designed activities fit into the larger school science curriculum.

Each column includes developmentally appropriate lessons for both grades K–3 and grades 4–6, thus allowing teachers to target their instruction to the appropriate grade level. Within each lesson, a purpose is provided to focus the learning, and a materials list and procedural outline are also included.

References and Internet Resources are often provided to direct teachers to supplemental materials and information.

Student pages are provided for many activities, especially those enhanced by having students record information or that require data to be collected.

Historical Use of Children's Literature in the Science Classroom

The use of children's literature for science instruction in the elementary classroom has been common practice for quite some time and has become an increasingly popular approach in recent years due to the need to maximize instructional time. Many teachers have that favorite go-to book that they use when teaching—some that immediately pop into mind include *A Drop in My Drink*, *The Salamander Room*, and *Call Me Ahnighito*. Students are able to elicit key scientific information from these books through effective reading strategies, thus mak-

Historical Use of Trade Books in the Science Classroom

ing reading science trade books a perfect way for students to build literacy skills while learning science content. Chapter 2 delves into the many different pedagogical reasons for the use of children's literature in the science classroom as well as several of the challenges teachers should consider when using this strategy.

Today's teachers have adopted, modified, tweaked, and even renamed some of the long-standing approaches to the use of trade books in the classroom. Throughout the past 40-plus years, teachers have incorporated children's literature into their classroom instruction. Some teachers have used books for read-alouds, others have displayed books for use as supplemental classroom materials, and still others have actually used trade books as the source of information.

One early source for identifying and encouraging the use of trade books in the science classroom is an initiative started by the American Association for the Advancement of Science (AAAS) in 1965. *Science Books and Films: Your Review Guide to Science Resources for All Ages*, which began as a print journal and is now available as an online subscription, focuses its efforts on print and nonprint materials in all of the science disciplines and for all age groups, K–college, as well as the general library audience.

Outstanding Science Trade Books for Students

The National Science Teachers Association has a long-standing history of identifying quality children's literature for use in the classroom and has provided teachers with a resource for selecting quality children's literature with accurate scientific content. Since 1973, NSTA has, with the Children's Book Council, jointly published an annual list of Outstanding Science Trade Books. The inaugural list first appeared in *Science and Children* and highlighted an annotated bibliography of 101 books that had been published in 1972. Since then, more than 2,500 books have been selected for inclusion in this list. Although the yearly selections are dependent upon the quality of the books submitted for consideration from publishing houses, books have included both fiction and nonfiction from many different genres, ranging from biography to poetry to picture books.

This annual list of Outstanding Science Trade Books originally focused on grades K–8; however, in 2002 the list was expanded to include books that could be useful for grades K–12. At that point, chapter books and longer, more in-depth informational books began appearing on the list. Books are currently categorized by archaeology, anthropology, and paleontology; biography; Earth and space science; environment and ecology; life science; physical science; and technology and engineering. Each annotation contains bibliographic data, publishing information, a brief synopsis of the book, intended grade-level audience, and notations for the alignment to the National Science Education Standards.

The selection criteria used by the panel of librarians, science educators, content specialists, and teachers to determine inclusion in the Outstanding Science Trade Books list is fairly straightforward. All books must be scientifically accurate, engage students in the understanding of science, and have an informative and aesthetically appealing format. Each book is read, text is scrutinized, and all illustrations and images are closely examined. As a result of these strong standards, only the highest quality of science children's literature is chosen. For example, the 1974 Outstanding Science Trade Book list included *Sunshine Makes the Seasons*, written by Franklyn M. Branley and published by Crowell for grades 2–4. The synopsis of the 34-page book read, "The illustrations by Shelly Freshman will definitely help children understand revolution and rotation and how the seasons occur. The type size and vocabulary make the reading easy and enjoyable" (NSTA 1975, p. 29).

This book, though no longer illustrated by Freshman and now published by HarperCollins, is in its third "reillustration" with the most recent re-release date of 2005. *Sunshine Makes the Seasons* is an example of a solid resource where the content

Historical Use of Trade Books in the Science Classroom

does not change even when the publishers decided to modernize the artwork to better appeal to today's reader.

The list of Outstanding Science Trade Books, however, serves another purpose for classroom teachers. Our understanding of scientific principles and phenomena are constantly changing—often at a much more rapid pace than schools or districts are able to keep up with by updating their science textbooks. Children's trade books are much less costly to replace than a textbook and easily supplement an outdated textbook chapter. Moreover, the Outstanding Science Trade Books list, a yearly review, recognizes when updates are made to reflect new scientific understanding in an otherwise beloved book.

In March 1988, for example, *Cancer: Can It Be Stopped?* by Alvin and Virginia B. Silverstein was included on the list. This book published in 1987 was released as an updated version of their 1977 book of the same title. Today, this book would be woefully outdated due to medical advances in the field of oncology and the new information that has been discovered since that point. In 2005, the Silversteins along with a third author, Laura Silverstein, wrote a more current book, appropriately titled *Cancer*, that was recognized on the Outstanding Science Trade Books List in 2006. The ability to have current and cutting-edge information available to children is essential for fostering their understanding that science is a dynamic and ever-changing field. Trade books, which are produced on a faster schedule than textbooks and sell at a much lower cost, can bring that new information into the classroom easily and serve as a springboard for investigations.

Science and Children's "Teaching Through Trade Books" Column

In the fall of 2002, a small group of science educators conceived, designed, and implemented a new column for NSTA's elementary-level journal, *Science and Children*, that would focus on the use of children's

literature to teach science. The reason for this new column was the growing use of children's literature in classrooms in general and in science classrooms in particular. In the later 1990s and early 2000s, the balanced literacy approach began to emerge and replace the whole-language approach as a force in the elementary classroom. The benefit of the balanced literacy approach at this time was that it taught "developmentally relevant literacy skills within the context of appropriately leveled reading materials of interest to the learner" (Reutzel and Cooter 1999, p. 3). Science content trade books could be identified by readability level, and even many books by well-known children's science authors were emerging as leveled readers. Take, for example, the second release of *Sunshine Makes the Seasons* in 1985. The edition included a notation that this book was a "stage two" book, which was intended for primary grades, in an attempt to market it to those teachers using leveled readers.

Beginning in January 2003, under the guidance of then field editor Dr. Chris Ohana, the column "Teaching Through Trade Books" first appeared in *Science and Children*—the inaugural column focused on snowflakes. For three and a half years, various teachers submitted original pieces on a variety of topics to column editor Christine Royce. In fact, 22 different authors contributed to the success of the column, demonstrating the broad appeal and popularity of the instructional approach of using trade books to help teach science.

In 2006, under the guidance of field editor Linda Froschauer, the column was turned over to veteran writers Karen Ansberry and Emily Morgan (who are the award-winning authors of *Picture-Perfect Science Lessons* and *More Picture-Perfect Science Lessons*) and Christine Royce (who has coauthored a series of books titled Investigate and Connect with David Wiley). Since then, the three have written the column in alternating months, always providing well-structured lessons with strong curricular connections to content that is developmentally appropriate according to the National Science Education Standards.

References

Conrad, P. 1995. *Call me Ahnighito*. New York: Harper Collins Publishers.

Hooper, M., and C. Coady. 1999. *A drop in my drink: The story of water on our planet*. London: Frances Lincoln Children's Books.

Mazer, A. 1994. *The salamander room*. New York: Dragonfly Books.

National Science Teachers Association (NSTA). 1975. Outstanding science trade books for children in 1974. *Science and Children* (12) 6: 27–30.

Reutzel, D. R., and R. B. Cooter Jr. 1999. *Balanced reading strategies and practices: Assessing and assisting readers with special needs*. Upper Saddle River, NJ: Prentice Hall.

Internet Resources

Children's Book Council
www.cbcbooks.org/readinglists.php?page=outstandingscience

Outstanding Science Trade Books
www.nsta.org/publications/ostb/?lid=tnav

Chapter 2

Why Use Trade Books to Teach Science?

We begin our teacher workshops by asking the simple question, "What was your favorite book as a child?" It is always fun to see faces light up and hear voices speak with excitement when teachers reminisce about their favorite books. Sometimes it's the pictures, rhyming verse, or humor that makes a book memorable. Other times, it's the characters, setting, or even our relationship with the person who read to us that makes us remember. But in more than 10 years of facilitating workshops, we have never heard anyone reply, "My fourth-grade science textbook was my favorite book as a child." Clearly textbooks have an important place in the science classroom, but using trade books to supplement a textbook or kit program can greatly enrich students' experience in the science classroom. Trade books engage us on an emotional level as well as an intellectual level. It's not just the words and pictures we remember; it's the feelings we associate with these books that stick with us.

So what exactly is a trade book? A trade book is simply a book intended for sale to the general public through booksellers, as opposed to a textbook or reference book. Most of the books featured in the lessons in this compendium are from a category of trade books known as picture books, which are "books in which the illustrations are of equal importance as or more important than the text in the creation of meaning" (Strickland and Morrow 2000). Reading experts Stephanie Harvey and Ann Goudvis (2007) claim that picture books top the list of literature that lends itself to reading comprehension strategy instruction, and we believe that this genre can also be one of the most effective for connecting kids with science content.

Engagement

One of the most compelling reasons to use picture books to teach science concepts is that picture books are highly engaging to students of all ages. According to Harvey and Goudvis, "Picture books, both fiction and nonfiction, are more likely to hold our attention and engage us than dry, formulaic text" (2007). Students are naturally interested in picture books, and we believe interest in a topic is a necessary precursor to understanding it. In today's standards-based world of education, we can no longer decide what we want to teach or have students decide what they want to learn. Our state standards and district curricula dictate which science concepts we are responsible for teaching students at each grade level. So, our job as teachers is to create interest in those concepts by making the subject matter compelling. This is often where the right picture book can work magic. For example, in Chapter 19, we use the book *Roller Coaster* by Marla Frazee to engage students in a lesson about force and motion. This wonderfully illustrated, humorous book about the high-interest topic of roller coasters draws students into a lesson about the position and motion of objects as outlined by the National Science Education Standards for grades K–4. Students can also be drawn in to a science lesson by a poignant storyline, such as the one in *Tiger Math* (Chapter 6). This book appeals to students on an emotional level as they hear the true story of TJ, an orphaned baby tiger born at the Denver Zoo, and the zoo staff's struggle to help him survive. The story engages students in a lesson on graphing as this dual-purpose book shows various ways the growth of the baby tiger can be displayed using a

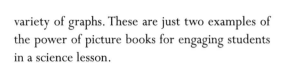

Why Use Trade Books to Teach Science?

variety of graphs. These are just two examples of the power of picture books for engaging students in a science lesson.

Context

Another key reason to use picture books to teach science is that they can give students more context for the concepts that they are learning. Some science concepts are either too abstract for students to easily understand or might not be very meaningful to students on a personal level. However, if the concept is part of a story that students can relate to, they will be able to better understand the concept. For example, a common objective taught in elementary school science is that day and night are caused by Earth's rotation. Students can memorize the statement, "Day and night are caused by earth's rotation," and even answer some test questions about the concept correctly, without having a deep understanding of this phenomenon. So, we use the book, *Somewhere in the World Right Now* by Stacy Schuett (Chapter 50), to give students a context for the concept. With beautiful, descriptive text, this book depicts several different scenes that could be happening somewhere in the world right now: Somewhere the sun is rising as people prepare for the day, somewhere people have their midday meal, somewhere the sun sets and shadows grow long, while somewhere else a child listens to a bedtime story. This book leads students to wonder, How can all of these things be happening at the same moment? Can the sun be "rising" and "setting" at the same time? Is it really "tomorrow" someplace else? *Somewhere in the World Right Now* creates a context for students to learn about Earth's rotation and gain a deeper understanding of its implications, including the need for different time zones.

Focus

Science textbooks can be overwhelming for many children and often cover a broad range of topics. While a page or two in a textbook can be helpful for teaching key vocabulary and providing explanations

for scientific phenomena, trade books can be used to dig deeper into a topic: "Unlike longer nonfiction or reference materials, picture books and other short texts focus our attention on one issue or topic at a time" (Harvey and Goudvis 2007). We suggest using trade books as supplements to a textbook or kit program to provide in-depth looks at particular science topics. For example, your science textbook might have a brief section in a chapter on force and motion about the forces of flight. In order to provide more depth on the topic, you could use the trade book *How People Learned to Fly* (Chapter 20). This book might reiterate the vocabulary, explanations, and diagrams found in the textbook—but in the more engaging context of the history of the human quest to fly.

Improved Reading and Science Skills

Research supports the use of trade books to teach science. Some studies have shown that students who are taught science using trade book literature as supplements to textbooks are better able to understand difficult scientific concepts and are more likely to read science-based books on their own (Moore and Moore 1989; Morrow, O'Connor, and Smith 1990). Other studies have shown gains in literacy as well as science achievement in programs that blend science and literacy instruction. Romance and Vitale (1992) found significant improvement in both science and reading scores of fourth graders when the basal reading program was replaced with readings in science that correlated with the curriculum. They also found that students' attitudes toward science improved.

Reading Aloud Has Many Benefits

We've all seen public service announcements urging parents to read aloud to their children. These aren't just suggestions to help parents bond with their children, they are based on research on reading success.

Why Use Trade Books to Teach Science?

In 1985, the findings of a landmark research study called Becoming a Nation of Readers (Anderson et al.) was released. In this report, the Commission on Reading found that "being read to is the most influential activity for building the knowledge for eventual success in reading." Reading aloud to children allows them to engage in texts with ideas above their reading level, and gives them support when encountering potentially challenging text features or vocabulary often found in informational texts. It also gives students opportunities to hear fluent, expressive reading and models the thinking strategies of a proficient reader. When a teacher does the reading, children's minds are free to anticipate, infer, connect, question, and comprehend (Calkins 2000). In addition, being read to is risk-free. Allen (2000, p.45) says, "For students who struggle with word-by-word reading, experiencing the whole story can finally give them a sense of the wonder and magic of a book."

Most of the lessons in this book involve the teacher reading the featured book aloud to students and providing them with opportunities to make connections to the text, talk to one another, jot down their thinking, and ask and answer questions. Harvey and Goudvis (2007, p. 48) refer to this experience as an "interactive read-aloud." They go on to explain, "The bottom line is that with this process, decoding doesn't interfere with understanding. All kids are free to listen, think about the ideas, talk to each other, and use strategies to understand the text."

The benefits of the read-aloud experience are not reserved for young children. We have found it to be effective for engaging students from preschool to high school. We believe reading aloud is appropriate in all grade levels and for all subjects. Allen (2000, p. 44) supports this view: "Given the body of research supporting the importance of read-aloud for modeling fluency, building background knowledge, and developing language acquisition, we should remind ourselves that those same benefits occur when we extend read-aloud beyond the early years. You may have to convince your students of the importance of

this practice, but after several engaging read-alouds they will be sold on the idea." In addition, reading aloud provides you the opportunity to model your own enthusiasm for reading, which can, in turn, provide encouragement for older students to want to read more. We like how Jim Trelease, author of *The Read Aloud Handbook,* likens read-alouds to a marketing strategy. He writes, "Every time we read aloud to a child or class, we're giving a commercial for the pleasures of reading " (2006, p. 37).

Trade Books Can Support Inquiry–Based Science

Trade books can be a valuable resource to integrate within an inquiry-based, constructivist approach to teaching science. Harvey and Goudvis write, "science trade books provide ample opportunity for children to ask and answer many of the questions they have about the natural world" (2007, p.69). Scientific inquiry is all about asking and answering questions. Sometimes trade books are a vehicle to inspire questions, and other times they are a resource for finding or checking answers. Morrison and Young (2008) suggest one approach in which trade books can support inquiry investigations: "If the students' explanations fit those presented in the book, the students can claim they have validated their findings. If they find inconsistencies between their explanations and those in the book, the students might be asked to propose reasons for the differences or to design a new investigation" (p. 208). Reading informational texts helps students "do science" as scientists do: designing investigations and comparing their results to what is already known.

Some of the lessons in this book are presented using the BSCS 5E Instructional Model (Bybee 1997). The phases of this model are Engage, Explore, Explain, Elaborate, and Evaluate. This learning cycle model is a way of organizing inquiry-based lessons in which students experience a scientific concept before they are provided with vocabulary or explanations. Then, they are given an opportu-

Figure 2.1. Phases of the 5E Instructional Model

Phase	Purpose
Engage	Spark interest and access prior knowledge.
Explore	Give students an experience with the concept.
Explain	Provide an opportunity for students to offer explanations of what they experienced in the explore phase and for the teacher to introduce vocabulary and refine student explanations.
Elaborate	Challenge students to apply new learning to a new situation.
Evaluate	Assess student understanding of the concept.

nity to apply what they have learned to a new, but similar, situation. The purpose of each phase of the 5E model is shown in Figure 2.1.

The biggest difference between the 5E model and traditional teaching is the order of the Explore and Explain phases. Traditionally, students are given vocabulary lists and readings before they have any experience with the topic. In the 5E model, the experience comes first. This creates a desire to know and provides a context for new vocabulary, readings, and explanations.

A Note About Misconceptions in Trade Books

We seldom do a workshop or speaking engagement without addressing the fact that some trade books can foster misconceptions in science. Be careful when selecting a book to share with students, making sure that it does not include inaccurate science. Children can get misconceptions from text—such as referring to all invertebrates as bugs (a term reserved for a specific group of insects)—as well as illustrations—such as diagrams that greatly exaggerate the Earth's elliptical orbit. Even though a book might look scientific, it could still contain inaccurate explanations. For example, we have found a number of trade books about plants that say plants take food from the soil. This is absolutely

incorrect and reinforces a common misconception that students have about plants. The idea that plants make their own food can be confusing to children, especially since "plant food" is sold at stores, but the truth is that plants make their own food through the process of photosynthesis. We do not use any of those books even though they might include other accurate information. Diana Rice (2002, p. 563) writes, "research has demonstrated that students learn not only 'good science' from trade books, but also encounter errors in their reading. As a result, teachers should exercise caution in selecting trade books for their science classes."

One helpful resource for selecting quality literature is a list of recommended trade books for teaching science that is released each year by NSTA in collaboration with the Children's Book Council. We like their criteria for selecting books:

- The book has substantial science content.
- Information is clear, accurate, and up-to-date.
- Theories and facts are clearly distinguished.
- Facts are not oversimplified so that the information is misleading.
- Generalizations are supported by facts and significant facts are not omitted.
- Books are free of gender, ethnic, and socioeconomic bias.

Why Use Trade Books to Teach Science?

It is important to note that many trade books were not intended to be used in the science classroom. Even though a book may look scientific, it is a good idea to check into the credentials of the author. We also suggest consulting with a knowledgeable colleague when selecting books. You might ask a high school teacher, college professor, or scientist to check a book for scientific accuracy. It is impossible to be an expert on every area of science, so don't be afraid to seek help when evaluating the quality of science trade books.

References

Allen, J. 2000. Yellow brick roads: *Shared and guided paths to independent reading* 4–12. Portland, ME: Stenhouse.

Anderson, R. C., E. H. Heibert, J. Scott, and I. A. G. Wilkinson. 1985. *Becoming a nation of readers: The report of the commission on reading.* Champaign, IL: Center for the Study of Reading.

Bybee, R. W. 1997. Achieving scientific literacy: *From purposes to practices.* Portsmouth, NH: Heinemann.

Calkins, L. M. 2000. *The art of teaching reading.* Boston: Pearson Allyn & Bacon.

Harvey, S., and A. Goudvis. 2007. *Strategies that work, second edition: Teaching comprehension for understanding and engagement.* Portland, ME: Stenhouse.

Moore, S. A., and D. W. Moore. 1989. *Literacy through content/Content through literacy.* Reading Teacher 43 (2): 170–171.

Morrison, J. A., and T. A. Young. 2008. Using science trade books to support inquiry in the elementary classroom. *Childhood Education* 84 (4): 204–208.

Morrow, L., E. O'Connor, and J. Smith. 1990. Effects of a story reading program on the literacy development of at-risk kindergarten children. *Journal of Reading Behavior* 22 (3): 255–275.

National Research Council (NRC). 1996. *National science education standards.* Washington, DC: National Academies Press.

National Science Teachers Association (NSTA). 2011. *Outstanding Science Trade Books for students K–12.* www.nsta.org/publications/ostb/?lid=tnav.

Rice, D. C. 2002. Using trade books in teaching elementary science: Facts and fallacies. *The Reading Teacher* 55 (6): 552-565.

Romance, N. R., and M. R. Vitale. 1992. A curriculum strategy that expands time for in-depth elementary science instruction by using science-based reading strategies: Effects of a year-long study in grade four. *Journal of Research in Science Teaching* 29 (6): 545–554.

Strickland, D. S., and L. M. Morrow, eds. 2000. *Beginning reading and writing.* New York: Teachers College Press.

Trelease, J. 2006. *The read aloud handbook, sixth edition.* New York: Penguin.

Children's Books Cited

Frazee, M. 2003. *Roller coaster.* Orlando, FL: Harcourt.

Hodgkins, F. 2007. *How people learned to fly.* New York: HarperCollins.

Nagda, A. W., and C. Bickell. 2000. *Tiger math: Learning to graph from a baby tiger.* New York: Henry Holt and Company.

Schuett, S. 1995. *Somewhere in the world right now.* New York: Alfred A. Knopf.

Chapter 3
Thought-Provoking Questions

By Christine Anne Royce

*W*hy, *what*, and *how*: Three words that young students often speak when they are full of questions about activities and experiences in their daily lives. Helping students clarify their thought processes and ask questions that can be answered through scientific inquiry is important. These skills will help them throughout their lives.

Trade Books

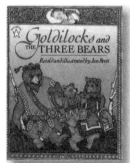

Goldilocks and the Three Bears
By Jan Brett
Puffin, 1996
ISBN 978-0-698-11358-9
Grades K–3

SYNOPSIS
Although this is my favorite version of this traditional tale, any will work for this activity. The three bears arrive at home to find that their chairs have been sat in, their porridge has been eaten, and someone was sleeping in baby bear's bed. The old standard "Who's been sitting in my chair?" leads to an answer that helps students realize that how you ask a question is as important as what you want to know.

June 29, 1999
By David Wiesner
Clarion Books, 1992
ISBN 978-0-395-59762-0
Grades 4–6

SYNOPSIS
A young girl launches seedlings into the atmosphere to determine what happens. What she doesn't expect are the giant vegetables that fall from the sky; she becomes puzzled about what happened and never finds out about the spaceship that jettisoned the vegetables. Although the story is implausible, students will realize that asking questions often leads to unexpected answers.

Curricular Connections

Students encounter a variety of experiences around which they formulate questions. Some of these can be answered easily by looking up the information (e.g., What is for lunch?), whereas others are more challenging and require greater investigation that may lead to a developmentally appropriate answer (e.g., Why doesn't the classroom pet like lettuce?). When students engage in simple investigations, they develop the ability to ask scientific questions, investigate aspects of the world around them, and use their observations to construct reasonable explanations for the questions posed (NRC 1996, p. 121). As students become familiar with this process, they will begin to understand that "scientists use different kinds of investigations depending on the questions they are trying to answer. Types of investigations include describing objects, events, and organisms; classifying them; and doing a fair test (experimenting)" (NRC 1996, p. 123).

Grades K–3: Considering the Question

Purpose

Students will learn to think about what information they have and ask questions that help them obtain more information to answer a puzzle.

Materials

- Chalkboard or chart paper
- Markers
- Ball
- Pencil
- Eraser
- Block
- Brown lunch bags
- Other common objects

Procedure

1. Read *Goldilocks and the Three Bears* and ask students to think about the questions the bears

ask and the answers they get. Students may see the pattern that the question asked provides a single-response answer. Some questions allow them to narrow down an answer one at a time. For example, the bears could have asked, "Who's been sitting in my chair?" and gotten answers such as the cat or the dog. Other questions help eliminate a series of possibilities, allowing them to arrive at an answer more quickly and more efficiently (e.g., Was the person sitting in my chair a boy or a girl?). The teacher will most likely need to model how to ask questions initially in this activity and then help students rephrase questions as they continue.

2. To conduct this activity, the teacher should have a series of objects that the students are familiar with, each placed in a brown paper bag or small box so the students cannot see the object or discern its shape. Begin with the first object—for example, a tennis ball—asking the students to develop a question that will help them learn the name of the object.

3. Students may begin by calling out random objects such as, "Is it a pen?" or "Is it a pencil?" Simply answer yes or no and record their guesses on the board. After several guesses, stop and ask: "Is simply guessing an effective method for finding out what the object is?" Some students may grasp the idea that it is a random approach. Modeling the "think-aloud" strategy, say, "Hmmm, if I were trying to figure out this object, what better type of question might help me narrow down what the object is?" (Accept student answers if they offer them.) Continue with the think-aloud strategy: "Perhaps, if I ask, 'Is the object something that is found in the classroom or outdoors?' I might be able to narrow down where the object might be located." Continue with this pattern of "Is the object something you find in a student's desk? Or on the teacher's desk?" Helping students see that using a certain type of question can narrow down what the object might be in this particular

Thought-Provoking Questions

activity models the desired questioning skill behavior. This may take several examples, but students can then model the behavior of asking clearer questions that help them arrive at the answer.

4. As students generate more productive questions, write them on chart paper or the board to revisit. After the game, ask students to look at their questions and consider how they helped narrow down the object. For example, "Do we use the object in the classroom?" would help students realize that the object might be used on the playground or at home (requiring another question to narrow the object further). Teachers can investigate other riddles that expand students' questioning skills one step further, such as Stories With Holes or other brainteasers (see Internet Resource).

Grades 4–6: Carefully Crafted Questions

Purpose

Students will ask a question that allows them to design an experiment or investigation.

Materials

- Vary based on the questions students develop

Procedure

1. Conduct a read-aloud of *June 29, 1999,* with the class. Stop at key points and ask the students to make observations about what is happening and predict what might happen. Ask, "What is Holly Evans doing with the seeds?" or, "What is a possible explanation for what happened when the vegetables start falling to Earth?" Students may believe that the seeds Holly launched grew into giant vegetables; however, they will get an unexpected answer. Explain that even when we ask a good question, we may get an unexpected answer. Students are often conditioned that they

must have the "right" answer and can be a bit timid about taking a risk and being incorrect.

2. Having the students use common, everyday items will allow them to focus on asking carefully crafted questions. Different types of questions lead to different types of investigations. Some will require students to look up information using reputable resources, whereas others will require them to design a simple investigation.

3. Ask, "If you could investigate any idea using the materials provided in the center of the room, what idea would you choose and why?" Provide materials such as magnets, toys, dice, coins, or balloons to investigate questions using these objects or other questions such as the ones listed here:

- What happens when you mix the primary colors of paint?
- What happens to your heartbeat when you jump rope for one minute?
- Where do pill bugs prefer to live?
- How does my breathing rate change when I exercise?

List all topics on the board and ask students to separate the questions into two categories—those we could easily explore in the classroom and those that would require us to go somewhere else. Ask the students to determine whether they have the materials within the classroom or if they would need to obtain or buy them elsewhere.

4. Continue to narrow down the question possibilities to arrive at a few that are "doable" and "reasonable to do" in the classroom. Identify possible investigations that could be done easily with the materials provided. The activities would depend on the age of the students and the curriculum areas they are studying, keeping it within the content areas of science.

5. Either the entire class can investigate the same question, which is recommended for younger students, or different student groups can choose individual questions to investigate. The latter may require assistance from classroom aides or parent volunteers. Regardless, the teacher will need to elicit what students' predictions are, the types of information they will gather, and what is the best way to record that information.

6. After they think through their experimental design process, allow teams to set up and conduct their investigations as well as collect data. Then have students consider their initial predictions—were they accurate or did they have to change their ideas due to the data they collected? Follow up by asking whether they generated any more questions by conducting their investigation. Some may have generated new questions based on their outcomes or what they observed during the activities—the process of inquiry!

Reference

National Research Council (NRC). 1996. *National science education standards.* Washington, DC: National Academies Press.

Internet Resource

Stories With Holes
www.storieswithholes.com/storwithol5.html

Chapter 4
A Closer Look

By Karen Ansberry and Emily Morgan

Give a child a hand lens or a microscope and they quickly become fascinated with the hidden worlds these tools reveal. These lessons provide opportunities for students to take a closer look at the properties of objects and organisms and to explore how magnifying instruments help scientists make observations and discoveries.

Trade Books

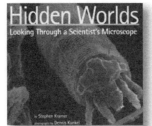

You Can Use a Magnifying Glass
By Wiley Blevins
Children's Press, 2004
ISBN 978-0-516-27328-0
Grades K–4

SYNOPSIS

Through simple text and photographs, this Rookie Read-About Science book describes how to use a magnifying glass and what kinds of things can be seen with it.

Hidden Worlds: Looking Through a Scientist's Microscope
By Stephen Kramer
Houghton Mifflin, 2001
ISBN 978-0-618-05546-3
Grades 4–8

SYNOPSIS

Hidden Worlds *takes you behind the scenes of scientist Dennis Kunkel's work and explains how he captures remarkable images of microscopic life and objects. Stephen Kramer's engaging text and Kunkel's dramatic photographs provide a fascinating look at a microscopist and the hidden worlds he explores.*

Curricular Connections

According to the *National Science Education Standards* (NRC 1996), students in kindergarten through grade 8 should learn how scientists depend on certain tools to help them make better observations. Instruments such as magnifying glasses and microscopes help scientists all over the world see, measure, and do things that they could not otherwise see, measure, and do. The Standards also highlight the importance of giving students opportunities to increase their understanding of the characteristics of objects and materials that they encounter daily. The lessons presented in this column provide students opportunities to observe, draw, and describe objects and organisms while teaching them how magnifying glasses or microscopes can enhance observations, drawings, and descriptions.

Grades K–3: With a Hand Lens

Materials

- Coin Observation student page (p. 19)
- Dimes (one per student)
- Hand lens
- Ink pad

Engage

Hand out the Coin Observation student page. Ask students to draw the features of a dime's "tails" side in detail, from memory, in the first circle. Have students share their drawings with one another, then discuss the limitations of drawing details from memory. Next, give each student a dime to observe with the naked eye. Have them use the second circle as an outline to draw the features of a dime (tails side again) in more detail. Then ask, "What scientific tool could you use to observe the details of an object better than you could with your eyes alone?" (e.g., magnifying glass or hand lens, microscope, binoculars, or telescope). Next, hold up a magnifying glass and tell students that it is a scientific tool that can help them see the features of an object in detail. Then, read aloud pages 3–11 of *You Can Use a Magnifying Glass.* Stop after reading *What can you see?*

Explore

A magnifying glass or plastic hand lens will help students get an even closer look at a dime. Model the proper way to use a magnifier (holding the lens close to one eye, shutting the other eye, and bringing the object toward the lens until it comes into focus), and caution them that touching things with the lens can scratch it. Then pass out plastic hand lenses to all students, and have them use the lenses to observe the backs of their hands, fingernails, pencils, and other objects. Some students may have trouble shutting one eye, so you may want to have them cover one eye with a hand or hold the hand lens out farther and use both eyes. After a few minutes of exploration and sharing, have students use the third circle as an outline to draw the tails side of a dime as seen through a hand lens.

Explain

Have students explain how the hand lenses helped them see details of things that they could not otherwise see. Then ask, "How does a hand lens work? What other things can you see with a hand lens?" Read the rest of *You Can Use a Magnifying Glass* and then discuss the answers to your questions. (As light hits a hand lens, the light rays bend. This makes things look bigger. You can see how a fly's eye is made up of lots of small pieces, you can look closely at jewels, and you can see the patterns in a fingerprint.)

Elaborate/Evaluate

Now students are going to use hand lenses to help them solve a mystery! In advance, prepare the "evidence." Have a fellow teacher use an ink pad to make a fingerprint at the top of a sheet of paper. Below that, have this person as well as several other teachers make fingerprints on the paper. Write their names below their respective prints. Then, explain

A Closer Look

to the class that something was borrowed from your classroom (e.g., a candy jar or a stapler) and you have collected a fingerprint from the object as evidence. You want to find out who borrowed the object, so you have collected fingerprints from some of the teachers in your school. But before they can solve the mystery, your students must first practice using hand lenses to study their own fingerprints. Reread pages 23–27 in *You Can Use a Magnifying Glass*. Have students use ink pads to carefully make their fingerprints, observe them with hand lenses, and compare them to other students' prints. Next, give each student or team of students a copy of the "evidence" and the "suspects." As they use their hand lenses to solve the mystery, observe whether or not they are using them correctly and safely. Then discuss how hand lenses can help people and scientists observe the properties of objects more closely.

Grades 3–6: With a Microscope

Materials

- SEM images
- Index cards or card stock
- Hole puncher
- Clear tape
- Newspaper
- Specimens to magnify (such as salt, sand, paper, fabric, and thread)
- Microscopes

Engage

In advance, print pictures of images taken with a scanning electron microscope (SEM). A good resource is the University of Hawaii's MicroAngela website: *www5.pbrc.hawaii.edu/microangela* (e.g., fly face, mosquito, or jumping spider face) Invite students to guess what each picture shows. After they guess, reveal what the object actually is, and ask them how they think the picture was taken. Tell students that the photos were taken with a special kind of microscope called a scanning electron microscope. Ask, "Who uses microscopes? What are they used for?"

Explore

We recommend low-powered microscopes (with a maximum magnification of 40× or less) for elementary students. Before doing the following activities, model how to properly operate and take care of the microscopes available for your classroom use. Tell students they are going to learn how to make their very own microscope slides. First demonstrate how to prepare simple, homemade slides by cutting index cards or cardstock into 1 in. × 3 in. strips, punching a hole in the center of each, and covering the hole with clear tape. Model how to mount specimens on the sticky side of the tape within the punched hole and label slides with the name of the specimen and the power of magnification. Next, have students make a slide of a lowercase *e* from a piece of newspaper and look at it through their microscopes. Ask, "Does the letter *e* look different through a microscope? How is it different?" (Students should notice that the letter *e* is not only larger and more detailed, but upside down and backward as well.) Allow students to create several slides of everyday items such as salt, sand, paper, fabric, and thread. Have them draw and label what they see in the microscope and write a few detailed observations about each specimen. Ask, "What kind of information can scientists gain from using a microscope? What kinds of scientists use microscopes?" Tell them that you have a book to help answer these questions.

Explain

Show students *Hidden Worlds: Looking Through a Scientist's Microscope* and ask them what they think is pictured on the cover. Tell students that it is a microscopic animal called a leaf gall mite that has been magnified 2,530 times using a scanning electron microscope. Next, read "A Note to the Reader" on page 7, which explains how the colors added to the images in the book were computer

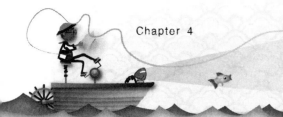

generated (and not necessarily the natural colors of the objects) and how the letters and numbers in parentheses throughout the book represent the type of microscope used and how many times the object was magnified. Next, read the text and captions on pages 8–13. Ask, "What are some of the ways that Dennis Kunkel has used microscopes for his own learning and to help other scientists?" (He has used microscopes to learn more about living and nonliving things that interest him and has helped scientists determine what meteorites are made of, identify a new species of ants, and figure out why spider silk is so strong.) Then ask, "How do you think Dennis got his start as a scientist?" and read pages 9–18 (second paragraph) about how Dennis began taking "collecting trips" as a child and later became a microscopist. Show the photos of the containers he uses for collecting on page 19, and then flip through the book to share some of the remarkable photos he has taken through a microscope. Next, announce that the class is going to take their own collecting trip to find specimens to examine with microscopes.

Check your district policy on taking students outside or away from the school grounds during the school day before doing the following activity.

Elaborate

Take students on a collecting trip to a natural area nearby. Provide small containers or film canisters with lids for collecting specimens in the field. Have each student collect three to five small items that will fit on a slide (no live animals), such as blades of grass, small pieces of tree bark, soil, seeds, and flower petals.

Evaluate

When you return to the classroom, evaluate students' abilities to make and label a handmade index card slide, operate a microscope properly, and draw and describe each magnified specimen in detail.

Reference

National Research Council (NRC). 1996. *National science education standards.* Washington, DC: National Academies Press.

A Closer Look

Name: _____

Coin Observation

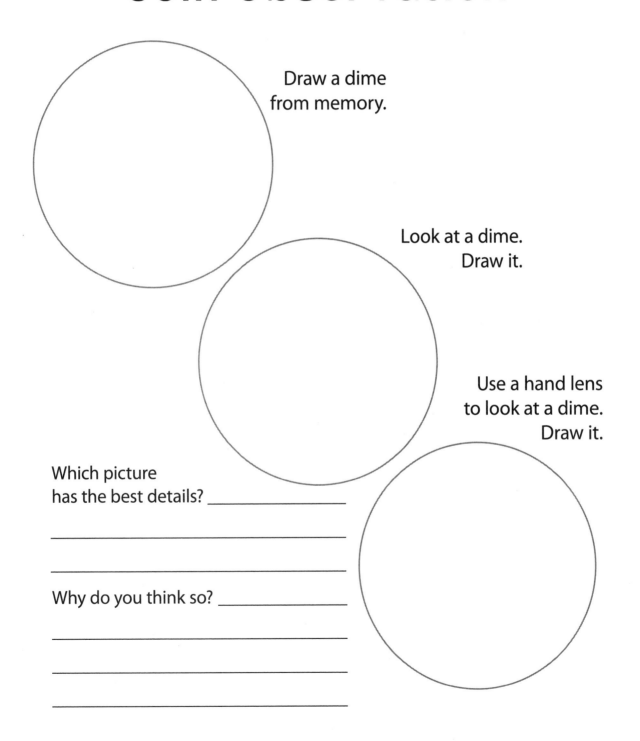

Draw a dime
from memory.

Look at a dime.
Draw it.

Use a hand lens
to look at a dime.
Draw it.

Which picture
has the best details? _____

_____

_____

Why do you think so? _____

_____

_____

Chapter 5

Science Measures Up

By Karen Ansberry and Emily Morgan

C an you measure a dog's tail in dog biscuits? Can you measure a desk without a ruler? Which is better: measuring a room in paces or meters? Which system of measurement do scientists use? Explore these questions and more to help learners understand why we use standard systems of measurement.

Trade Books

Measuring Penny
By Loreen Leedy
Henry Holt and Company, 2000
ISBN 978-0-8050-6572-5
Grades K–4

SYNOPSIS
Lisa learns about standard and nonstandard units of measurement by measuring her dog Penny with all sorts of units, including pounds, inches, dog biscuits, and cotton swabs.

How Tall, How Short, How Faraway
By David A. Adler,
illustrated by
Nancy Tobin
Holiday House, 2000
ISBN 978-0-8234-1632-5
Grades K–4

SYNOPSIS
Simple text and cartoonlike illustrations introduce the history of measurement systems, beginning in ancient Egypt and ending with the modern metric system.

Curricular Connections

The Science as Inquiry standard of the *National Science Education Standards* (NRC 1996) includes measurement as a fundamental ability necessary to do scientific inquiry. Students should be able to employ simple equipment and tools to gather data and extend the senses. *The National Science Education Standards* also suggest that children develop some essential understandings about science and technology, including the idea that people throughout history have invented tools and techniques to solve their problems.

Weights and measures were among the first tools invented by man. Ancient people used their body parts and items in their surroundings as their first measuring tools. As societies evolved, measurements became more complex. By the 18th century, England had achieved a greater degree of standardization in measurement than other European countries. The English, or customary system of measurement commonly used in the United States, is nearly the same as that brought by the colonists from England.

The need for a single, worldwide measurement system was recognized in 1670, when a French priest named Gabriel Mouton proposed a measurement system (based on units of 10) that was both simple and scientific. However, a century passed and no action was taken. During the political upheaval of the French revolution in the 1790s, the French Academy of Sciences proposed a new system, based upon Mouton's, as a way to bring order to the confusing and often contradictory traditional systems of weights and measures that were being used throughout Europe. The metric system got its name from the unit of length, called a *meter,* which is derived from the Greek word meaning "a measure." The standardized structure and decimal features of the metric system made it well suited for scientific and engineering work, and wide acceptance of the metric system coincided with an age of rapid technological development. Although the English system of measurement is commonly used in everyday situations in the United States, scientists around the world primarily use the metric system (known as SI, from the French Systeme Internationale d'Unites) in their daily work.

Grades K–3: Measuring Pets

Materials

- Stuffed animals
- Measuring tapes
- Everyday items that can be used as nonstandard units of measurement (e.g., cotton swabs, dog biscuits, and paper clips)
- Duel-sided rulers
- Meterstick
- Measuring Pets student page (p. 25)

Engage

Ask, "Can you measure a dog's tail in dog biscuits?" Then show students the cover of the book *Measuring Penny* and explain that in this book, Lisa measures her dog Penny in a variety of ways. Make connections by asking students to share their own experiences with measuring, and then read *Measuring Penny* aloud to the class. Pause after reading pages 7 and 8, where Mr. Jayson gives the class "Measuring Homework," and point out that there are two parts to a measurement, a number and a unit. Ask students to signal each time they hear an example of a different unit of measurement as you read the rest of the book aloud. After reading, revisit page 8 where Mr. Jayson gives examples of standard and nonstandard units and ask, "What is the difference between standard and nonstandard units?" Students should realize that *standard units* are units of measurement that are accepted and used by many people and *nonstandard units* are everyday objects that can be used for measuring.

Explore/Explain

Have students bring in a favorite stuffed animal to measure in both standard and nonstandard units.

Science Measures Up

Provide measuring tapes as well as various items they could use as nonstandard units (e.g., cotton swabs, dog biscuits, and paper clips). After some modeling and guided practice, have students measure the length of their stuffed animals' parts in as many ways as they can using a standard unit and a nonstandard unit for each part. Students should record their results on the Measuring Pets student page (p. 25). As students are measuring, circulate to ask the children how they arrived at their measurements and to explain how they classified each unit as standard or nonstandard. Then have them trade stuffed animals with a partner and check one another's measurements.

Elaborate/Evaluate

Ask, "Which units do you think are best for making accurate measurements, standard or nonstandard? Why?" Students should be able to explain that standard units are best because they are always the same; for example, an inch is always the same length but dog biscuits can be different lengths. Explain that most people around the world, as well as scientists, use a standard system of measurement called the metric system because it is simpler and more scientific than the English (or inch-pound) system.

The metric system was invented more than 200 years ago, but people in the United States have not entirely switched over to it. Have students find the metric side of their rulers and point to a centimeter. Explain that a centimeter is about the width of a pinky finger. Then show students a meterstick and explain that it is about as long as their outstretched arms. Have them find something in the room that is about a centimeter long and something that is about a meter long. Then have them measure their desks in centimeters and the length of the classroom in meters. Evaluate their understandings about metric measurement by asking questions such as, "Why do we need standard units of measurement? What standard system of measurement do scientists use? Which metric unit would be best for measuring the length of a dog's tail? The length of a school bus? A book?" and so on.

Grades 4–6: History of Measurement

Materials

- Metersticks

Engage

Ask, "Can you measure a desk without a ruler?" and then challenge students to measure the length of your desk without using any traditional measuring tools. As a class, brainstorm a list of ways that you could measure the desk. Tell students that in ancient times, measurement tools were not readily available, so people had to come up with creative ways to measure things. In ancient Egypt, one way to measure was the "span." A *span* is from the tip of the thumb to the tip of the little finger with the hand stretched wide. Have a student measure your desk with his or her hand span and record the number of spans on the board. Call on several other students to measure the desk using their hand spans and record the number of spans on the board each time. Then measure the desk using your own hand span, and record that number of spans on the board. Ask, "Which measurement is the correct answer for the length of the desk?" Students should understand that there is no "correct" answer in spans because each person's span is a different size.

Explore/Explain

Introduce the book *How Tall, How Short, How Faraway,* and then read through page 7 about how the hands, fingers, and arms were used as measuring tools in ancient Egypt. Have students try measuring their heights using ancient Egyptian nonstandard units (cubits, spans, palms, and digits) as explained on pages 6–8. Next, tell students to have their heights measured by partners using cubits, spans, palms, and digits and compare those measurements to their own. Ask, "Is this an accurate way to measure? Why or why not?" Read the rest of the book aloud, and then ask, "Which metric unit would you use to measure your height?" Have students measure their

own heights in centimeters and compare them to their heights measured in centimeters by partners. Ask, "Is this an accurate way to measure? Why or why not?" Then ask, "Which is better: measuring a room in paces or meters?" Have students measure their classroom in both paces and meters and then explain which method they think is more accurate and why.

Elaborate

Tell students that the standard system of measurement used by most countries of the world—and by scientists everywhere—is the metric system. Have students do research to create a Metric Measurement poster that includes: a timeline describing and illustrating the major events in the development of the metric system, a table showing four or more common metric units and an example of something that might be measured with each, and an argument for or against adopting the metric system for all measurements in the United States. For fun, students can include a song, rap, or cheer promoting the use of either the metric system or the customary system.

Evaluate

Have students present their metric measurement posters to their classmates. In the presentations, they should include why they chose to include certain events in their timelines, their metric unit examples, and their arguments (and song, rap, or cheer) for or against adopting the metric system for all units of measurement in the United States.

Reference

National Research Council (NRC). 1996. *National science education standards*. Washington DC: National Academies Press.

Internet Resources

Metric History Timeline
http://library.thinkquest.org/J002831/metrictimeline. htm

U.S. Metric Association
http://lamar.colostate.edu/~hillger

Science Measures Up

Name: _____

Measuring Pets

Your Assignment:

1. Choose a stuffed animal to measure.

2. Measure the length of its parts in as many ways as you can. Use a standard and a nonstandard unit for each part you measure. Be creative!

3. Record your results.

Remember, a measurement always has two parts:

1. a number

2. a unit

Example:

ANIMAL PART	STANDARD UNITS	NONSTANDARD UNITS
ear	12 centimeters	1½ cotton swabs

My Animal

ANIMAL PART	STANDARD UNITS	NONSTANDARD UNITS

After you measure, draw a picture of your animal on the back. Label the parts you measured.

Chapter 6

Going Wild With Graphs

By Karen Ansberry and Emily Morgan

Just as words can tell a story, so can graphs. Pick up a newspaper or magazine and you will probably see several articles accompanied by graphs. Graphs are useful because they communicate information visually and can usually be read more quickly than the raw data from which they are made. Use the high-interest topic of zoo animals to teach students about using graphs to organize and communicate data.

Trade Books

Giraffe Graphs

By Melissa Stewart
Children's Press, 2007
ISBN 978-0-516-24594-5
Grades K–4

SYNOPSIS

In this Rookie Read-About Math book, a group of students take a field trip to the zoo, where they learn how to tally numbers and create graphs. Colorful photos and simple text introduce students to graphing.

Tiger Math: Learning to Graph From a Baby Tiger

By Ann Whitehead Nagda and Cindy Bickel
Owlet Paperbacks, 2002
ISBN 978-0-8050-7161-0
Grades K–4

SYNOPSIS

Readers are introduced to picture graphs, circle graphs, bar graphs, and line graphs—all which help tell the story of T. J., an orphaned Siberian tiger cub born at the Denver Zoo.

Curricular Connections

One of the essential tools of scientific inquiry is the use of mathematics. The *National Science Education Standards* (NRC 1996) state that mathematics can be used by students to ask questions; to gather, organize, and present data; and to structure convincing explanations. The lessons here feature the mathematical skills of reading, analyzing, and creating graphs, and they show students how these skills are useful in science.

In the K–2 lesson, students work with an adult helper to find a graph in a magazine, newspaper, or internet article to share the next day. Students then create their own tally chart and bar graph representing their favorite animal pictured in the book. In the 3–6 lesson, students also bring in graphs from newspapers, magazines, or internet articles, which they classify as picture graphs, bar graphs, line graphs, and circle graphs. They also explain when it is appropriate to use each of these types of graphs. The lesson culminates with students creating their own graphs based on real-life data either shared by a zookeeper or naturalist or collected from observations and measurements of a class pet.

Grades K–2: Favorite Zoo Animal Graphs

Materials

- Scissors (at home)
- Magazines (at home)
- Favorite Zoo Animals student page (p. 31)

Engage

Show students the cover of *Giraffe Graphs*. Read aloud the first page of the book and then encourage connections by discussing the following questions: Have you ever been to a zoo? What types of animals can be found in a zoo? What things could you count at the zoo? What things could you measure at the zoo? Read the book aloud, providing time for students to answer the questions posed in the book. In addition, have students answer questions about the tally charts and bar graphs found in the book such as, "What is being counted in this tally chart?" and "What does this graph show?"

Explore

Giraffe Graphs shows students how bar graphs can be used to share information. The book ends with this challenge: "When I go home, I'm going to look for graphs in magazines and in the newspaper. Why don't you look for some graphs, too?" For homework, have students and their adult helpers locate graphs in magazines, newspapers, or internet articles and cut them out to share at school the next day. Remind students not to cut off the graph title and labels, because they help the reader understand what the graph is showing.

Explain

Give students time to share the graphs they have collected with the class. Ask them to read the title and the labels and to explain what they think the graph is showing. For example, "My graph is called Daily Temperatures. It has the days of the week along the bottom of the graph and temperature along the side of the graph. It shows how this week's temperature changed from day to day."

Next, have students make a display in the classroom using the graphs.

Elaborate

Refer back to *Giraffe Graphs* and ask students if they can remember all of the animals that were shown in the book. Flip through the book to find pictures of zebras, giraffes, elephants, and lions. Pass out the Favorite Zoo Animals student page (p. 31) and invite students to think about which of the four animals in the book is their favorite. Have students raise their hands to vote for their favorite and count the votes for each animal. Next, have students record the number of votes for each animal using tally marks on the Favorite Zoo Animals Tally Chart. Students can add up the tally marks and record the sums in the "Total" column and then color in the Favorite

Going Wild With Graphs

Zoo Animals Bar Graph to represent the totals. Next, provide students with a variety of zoo animal pictures or small plastic zoo animals. Have them use these items to practice making tally charts and bar graphs by counting and graphing kinds of animals, number of animals with spots, number of animals with stripes, number of carnivores, number of herbivores, and so on.

Evaluate

Using their Favorite Zoo Animals Bar Graph, have students answer such questions as, What does this graph tell you? (The number of votes each animal received.) Which animal was the class favorite? How do you know? (It has the highest bar.) Which animal was the least favorite? How do you know? (It has the lowest bar.) How many more votes did the _____ get than the _____? and so on.

Grades 3–6: Types of Graphs

Materials

- Colored pencils or markers
- Poster paper

Engage

Introduce the book *Tiger Math* to students and read the introduction on page 7. Tell students that just as the introduction suggests, the first time through you will only read the right-hand pages, which tell the story of T. J., a Siberian tiger. As you read T. J.'s story, have students interact with the text by discussing questions such as, What do you think will happen to T. J.? (p. 10), How do you think the zoo staff will get him to eat? (p. 17), Have you ever seen kittens or puppies play this way? (p. 25). The focus of this first day should be sparking student interest in the story. Tell students that later you will read the graphs on the left-hand pages to learn more and to see exactly how T. J. grew.

Explore

For homework, have students search for graphs in magazines, newspapers, or internet articles and answer the following questions about the graphs they found: Where did you find the graph? What is the title of the graph? What are the labels on the graph? What is being counted, measured, or compared? What conclusions can you make from reading the graph? Give students the opportunity to share their graphs and their answers to these questions.

Explain

Read aloud the left-hand pages of *Tiger Math* featuring the graphs. Point out the different types of graphs and the features of each one (e.g., title, labels, and numbers). Ask students to draw conclusions from each graph as you read, such as "There are more Bengal tigers in the wild than any kind of tiger" (p. 8) and "T. J.'s weight increased as he got older" (p. 14).

After reading, ask students if they can recall the four types of graphs shown in *Tiger Math* (picture graphs, bar graphs, line graphs, and circle graphs). Label the board with these five categories: Picture Graphs, Bar Graphs, Line Graphs, Circle Graphs, and Other. Then have students classify the graphs they have collected into these categories. Explain to students that picture graphs and bar graphs are generally used to compare different amounts (such as number of kinds of tigers in the wild on p. 8), line graphs are typically used to show changes over time (such as the change in T. J.'s weight over 12 weeks on p. 18), and circle graphs are typically used to show parts of a whole (such as the percentage of each species of tiger on p. 10).

Elaborate

Ideally, students should create graphs using real-life data on an animal. Contact a zoo or nature center and ask if someone there can share data with your class about an animal's weight, height, eating habits, behaviors, and so on. Or, you can have students collect data on classroom pets by measuring the animals and/or observing their activities each day. Be sure to check your school's policy on animals in the classroom. Once data are collected on the animals, students can decide how best to represent that data using graphs.

Evaluate

Students can create graphs using the data, explain why they created that type of graph, and share their graphs with the rest of the class in a poster session. Provide colored pencils or markers so students can use color on their graphs. A good online resource for helping students create bar graphs, line graphs, and circle graphs is Create a Graph from the National Center for Educational Statistics: *http://*

nces.ed.gov/nceskids/createagraph/default.aspx. This website provides information on how to choose the right type of graph for your data. It also allows students to enter their data and create professional-looking graphs that they can print.

Reference

National Research Council (NRC). 1996. *National science education standards.* Washington, DC: National Academies Press.

Internet Resource

Create a Graph
http://nces.ed.gov/nceskids/createagraph/default. aspx

Going Wild With Graphs

Name: _____

Favorite Zoo Animals

Favorite Zoo Animals Tally Chart		
ANIMAL	**TALLY**	**TOTAL**
Giraffe		
Elephant		
Zebra		
Lion		

FAVORITE ZOO ANIMALS BAR GRAPH

15
14
13
12
11
10
9
8
# of Votes 7
6
5
4
3
2
1
0

Giraffe   Elephant   Zebra   Lion

Animals

Chapter 7

Wild About Data

By Emily Morgan and Karen Ansberry

*I*n today's world of short attention spans and media overload, the ability to create visual representations to communicate data is an important skill. These lessons focus on various ways to display data and the purpose of visual representations when trying to communicate data to others.

Trade Books

The Great Graph Contest
By Loreen Leedy
Holiday House, 2005
ISBN 078-0-8234-1710-0
Grades K–3

SYNOPSIS
Gonk and Chester, two amphibian friends, hold a contest to see who can make better graphs. Loreen Leedy's fun illustrations show various types of data representation, such as circle graphs, pictographs, and bar graphs.

Collecting Data in Animal Investigations
By Diana Noonan
Capstone Press, 2010
ISBN 978-1-4296-5237-7
Grades 3–6

SYNOPSIS
This book from Capstone's Real World Math series follows Mr. Martin's class as it prepares for a field trip to the city park. Each student is responsible for collecting and organizing data about an animal that can be found at the park.

Curricular Connections

The *National Science Education Standards* (NRC 1996) stress the importance of using appropriate techniques to collect, organize, and communicate data in all areas of scientific study. In the early grades, students should be provided opportunities to gather data and use data to construct reasonable explanations. This month's lesson for K–3 involves collecting data to answer a question and organizing the data using a graph. Students learn that the type of data to be collected and the type of graph to be used depend on the question being asked. They also learn that tables are a simple way of organizing the data they collect. In the "Elaborate" portion of the lesson, the students choose a question to investigate, decide what data to collect, and create a graph to display the data. In the upper elementary grades, the Standards suggest that mathematics is essential to asking and answering questions about the natural world. Students should use mathematics to ask questions, gather data, organize and present data, and structure convincing explanations. In the lesson for grades 3–6, students collect data about various animals in their local environment. They learn the difference between quantitative and qualitative data and how both types are important. Through examples in a read-aloud, students also learn various ways to display and communicate data. Last, with the help of the teacher, students come up with engaging ways to display their animal research data on posters to share with the rest of the class.

Grades K–3: The Great Graph Contest

Materials

- Sticky notes
- Chart paper
- Markers

Engage

Begin by asking a question to students that is relevant to them and related to the content they are learning. If you are beginning a unit on animals, you might ask, "If you were a scientist who studies animals, which animal would you want to study?" If you are beginning a unit on health, you might ask, "What is your favorite exercise activity?" Have the class suggest some answers and narrow them down to five or six.

Explore

Ask each student to write his or her answer to your question on a sticky note and place it randomly on a piece of chart paper. Read through some of the answers aloud. Ask students for a better way to organize their responses to make it quicker and easier to determine how many students gave each answer. Next, write the choices across the bottom of the board or chart paper and stack the corresponding sticky notes on top. Ask students whether that made it easier to compare the number of answers. Then ask them what they could add so that someone else that came into the room would know what this poster was about. Add labels and a title. Tell students what they just made is called a *bar graph*. Then show them the cover of *The Great Graph Contest*.

Explain

As you read, point out how the information was collected and then organized in each type of graph. Also, explain that each graph was created to answer a question. Read the question for each graph, such as "Are there more chocolate chip, oatmeal, or sugar cookies?" on pages 12 and 13. Then ask students to use the graph to determine the answer. They should notice that the stack of chocolate chip cookies is the tallest, so that bar contains the most cookies. Explain that graphs make it quicker and easier to answer questions. You can usually tell the answer at a glance. Read the section at the back of the book titled "More About Graphs," which explains how each graph was used to represent a set of data.

Wild About Data

Explain that *data* means "information that is collected." For example, on page 31, the answers to the survey on the clipboard are the data that Beezy collected. Tell students that these responses were collected using a data table. Point out the labels on the table and discuss how creating a table makes it easier to collect data. Then, show them how Beezy used the data in the table to create a graph.

Elaborate

Tell students that the class is going to have a "Great Graph Contest." Each pair will enter a graph in the contest. The instructions for making a graph can be found on page 32. They are (1) think of a question to ask; (2) collect data and write it down; (3) use real objects or art supplies to create your graphs; and (4) put titles and labels on your graphs so people will know what they are about. Encourage students to create data tables before collecting their data. Check student's work at each step before they move on to the next one.

Evaluate

Have students display their graphs in the classroom or the hallway. Invite another class to vote for the best graph by writing down votes on sticky notes. Create a sticky note bar graph on the board to display the contest results and determine the winner.

Grades 4–6: Displaying Animal Data

Materials

- Internet
- Nonfiction books about local animals
- Poster paper
- Markers
- Sticky notes

Engage

Have students help you generate a list of 30 or more native animal species from your area. Include fish, amphibians, reptiles, birds, mammals, invertebrates such as spiders and insects—and have students be specific (e.g., instead of "woodpecker" list "downy woodpecker"). Have students search your state's Department of Wildlife website or the National Fish and Wildlife Service website (see Internet Resource) for more species, or have them look through books about local wildlife.

Explore

Have each pair of students choose one of the local animals on the list to research. Ask them to collect information to answer questions such as, Where would you find this animal? What does it eat? How long does it live? Information that they collect can be through internet or library research or based on their own experiences and observations. Give students several days to collect information and have them share what they have learned and how they recorded the information (e.g., lists, paragraphs, or pictures).

Explain

Show students the cover of *Collecting Data in Animal Investigations*. Read pages 4 and 5 and stop when you get to the bold print word *data*. Model how to use a glossary by looking up *data* in the back of the book. Then, explain that there are two types of data: quantitative and qualitative. Quantitative data involves quantities, or numbers. Qualitative data does not involve numbers. While reading pages 6 and 7, stop to point out the quantitative data about the dragonfly (e.g., three body parts) and some of the qualitative data (e.g., labeled diagram of dragonfly). As you read, stop and ask students to identify different examples of each. Also, discuss the various ways in which the students in the book display their data (graphs, tables, lists, photographs, diagrams). Have students practice different ways to display data, creating some of the tables and graphs described in the "Let's Explore Math" insets throughout the book. Point out the differences between a table and a graph. Explain that sometimes, a table is the simplest way to display data,

such as the "Mallard Duck Eggs" table on page 9. Next, have students use highlighters to identify the quantitative data they collected about their animals. If students have more qualitative than quantitative data, ask them what quantitative data they could add to their research.

Elaborate

Tell students that each pair will create a poster that teaches the rest of the class about the animal they have been studying. Before they begin, point out the different ways the students in the book displayed their data. Ask your students why they think the students in the book decided to share their data in these forms. Tell students that their challenge is to create a poster that uses visuals to display the data they have collected about their animals. They will be evaluated on the accuracy of the information in their posters and the way the data are organized and displayed. Ask them to think of ways to display the data that would make the poster interesting and easy to read.

Evaluate

Have a gallery walk where pairs display all of the posters in the classroom. Give each student a pack of sticky notes and have them write a comment or question about the featured animal on one sticky note and a comment or suggestion about how the data are displayed on another sticky note. The sticky notes can be placed on or near each poster for the pairs to read through after the gallery walk.

Reference

National Research Council (NRC). 1996. *National science education standards*. Washington, DC: National Academies Press.

Internet Resource

U.S. Fish and Wildlife Service Digital Library
www.fws.gov/digitalmedia

Chapter 8

Taking Note of Natural Resources

By Christine Anne Royce

T he idea of "going green" or being aware of and reducing our impact on natural resources is receiving a lot of attention these days. Schools are starting to employ green practices and are soliciting help from students to be aware of and reduce their environmental impact. The trade books chosen for these investigations examine the use of natural resources from two different perspectives—what we do with them and how they are depleted over time.

Trade Books

Weslandia
By Paul Fleischman
Scholastic, 2000
ISBN 978-0-0439-22777-3
Grades K–3

SYNOPSIS
A young boy decides to develop a civilization of his own by growing a particular plant. All aspects of the civilization must come from what this plant produces. Through his discoveries, he examines different needs of the civilization from food to clothing to shelter and how the availability of resources affects the production of these items.

Just a Dream
By Chris Van Allsburg
Sandpiper, 2011
ISBN 978-0-547-52026-1
Grades 3–6

SYNOPSIS
Chris Van Allsburg uses illustrations and narrative text to introduce the reader to the idea that they affect their environment. Through Walter's adventures when he falls asleep, he explores how resources have been depleted and how individuals can change their actions to protect the environment.

Curricular Connections

What is your environmental impact? The idea that each human impacts or leaves a footprint on the environment based on the amount and type of natural resources used connects to the science curriculum through biology, chemistry, Earth science, and more. The idea of natural resources is presented in an important but often forgotten section of the *National Science Education Standards* (NSES 1996)—Content Standard F: Science in Personal and Social Perspectives. The *NSES* describe resources as "things that we get from the living and nonliving environment to meet the needs and wants of a population [and include] basic materials, such as air, water, and soil; [as well as] some [that] are produced from basic resources, such as food, fuel, and building materials" (NRC, 1996, p. 140).

Young students are often able to discuss the idea of pollution (the end product of natural resource use) but often have difficulty thinking about what resources go into making different items. By reading *Weslandia*, students will have the opportunity to think about what goes into the different products they use. They can then examine how many of their items come from or contain natural resources. Older students can engage in this same activity but go one step further to examine how humans' use of these resources affects their availability over time through a mining simulation.

Grades K–3: What Are Objects Made Of?

Purpose

Students will classify objects as living or nonliving and investigate what natural resources make up common household items.

Materials

- Pencil
- Toothpaste
- Piece of paper
- Picture of a sidewalk
- Drinking glass
- Fruit or vegetable
- Plant
- Informational handouts (See Internet Resource.)

Procedure

1. Read *Weslandia* to the class. Afterward, ask students to describe what they think the main idea of the story is. Students will talk about the civilization that Wesley makes. Reread the book to the class, this time focusing on the different items that Wesley needs for his civilization to be successful—food, clothing, shelter, transportation, and so forth.

2. After the second reading, ask students to brainstorm a list of items that they use every day that they "must" have to survive. Many of the items students name (e.g., TV, iPods, computers) are not necessary for survival, but in their minds they are. Once a list is generated, introduce the idea that items fall into one of two categories—living or nonliving. After reviewing the characteristics of living and nonliving things (e.g., living things need food and water and to be able to make new living things like seeds and babies; nonliving things don't need these things), introduce the idea of renewable and nonrenewable resources. Living items are capable of reproducing and are therefore renewable and able to be replenished—provided we practice sound management. Nonrenewable resources, however, are those that we cannot get more of, or replenish, because these resources take a very, very, very long time or are impossible to remake. In this scenario, living items are renewable: We can plant more crops and breed more animals, and the human population is growing. However, many nonliving items come from nonrenewable natural resources (e.g., oil, gas, metals).

Taking Note of Natural Resources

3. After this discussion, call students' attention to the objects you brought to class. Ask students to think about what each item is made of, then share with them information about the various objects from the handouts. Students will likely be surprised that pencils contain both tree wood (a living, renewable resource) and graphite (a nonliving, nonrenewable resource). The idea that many different types of resources may make up a common item is surprising to students—in fact they probably have never thought about what makes up many of the items they use every day. Then, ask students to classify each object as living or nonliving and to think about whether it is made from renewable or nonrenewable resources. Most objects that are food come from living, renewable objects, whereas nonfood objects come from nonliving or living resources.

4. After students have had a chance to think about the natural resources that go into common items that are used daily, revisit their list of things they need to survive. Ask them to determine if the items could have been made from the plant that Wesley uses in Weslandia and to explain their reasoning—the electronics they likely named will not be able to be made since they are not composed of renewable natural resources!

Grades 4–6: Pasta Mining

Purpose

This simulation activity helps students understand how natural resources are depleted.

Materials

- 1 lb. of six different types of pasta of varying size and shape
- Large bedsheet or drop cloth
- Transparency graphs
- Stopwatch
- Mining Natural Resources student page (p. 41)

Procedure

1. As you read *Just a Dream* to students, have them stop and discuss what they are "hearing" in the words as well as what they are "seeing" in the pictures. Ask, "What is happening to the natural resources in Walter's world? Do you think this happens today? Why or why not?" Many students will respond that we do need resources, but it is not as bad as Walter experiences. Students often don't equate what they personally use to the overall world problem of dwindling resources.

2. Next, introduce the activity on pasta mining. The premise of the activity is that there will be six different mining companies whose job it is to "mine" or "extract" their particular resource— in this case, a type of pasta.

3. Each mining company of five or so students can only mine their assigned resource (a type of pasta) in shifts, and they cannot destroy other resources (the other types of pasta). This would be a great time to discuss some of the restrictions associated with mining today, thus connecting to the societal perspective of the NSES. Mining companies must secure permits for the type of resource they want to mine; plans must be in place to assure that land reclamation will happen after the mining ceases; and there are many environmental laws associated with the process relating to land, water, and air pollution.

4. The mine is a large drop cloth or bedsheet on which the various pastas have been mixed, then spread.

5. Have each group set up a data table in which to record their results. In timed shifts of one minute, students will take turns mining as many pieces of their resources as possible without destroying others. The students should pick up only their resources and place them in a pile before returning to their team when the time is up. The process is repeated for the number of students in the group.

6. When all the trials are complete, the group develops a bar graph representing the amount of resource collected versus the trial number.

7. As this activity is conducted, several things may happen. Students may become competitive about the amount of resource mined, which also happens in the mining industry due to the limited amount of identified and located resources. The companies with larger types of pasta usually can collect their resource earlier, as larger pieces are more visible and easier to find, whereas the smaller pastas are usually more abundant in the later trials because they are easier to find after some of the other resources have been cleared away. To real-world mining companies, the size of the resource and ease of extraction will affect when and how much can be mined.

8. As students display their graphs, discuss what happened during the mining process and speculate possible reasons for success (or lack thereof) at collecting their resource. Ask, "How did resource availability change over time? Does the type of resource affect your ability to mine it? Would it have helped if companies could have worked together? What might happen to the resources over time?" Most students conclude that nonrenewable resources are depleted over time depending on the need for the resource and the rate of mining.

Reference

National Research Council (NRC). 1996. National science education standards. Washington, DC: National Academies Press.

Internet Resource

Mineral Information Institute Handouts
www.mii.org/pdfs/Digs_Color_8in.pdf

Mining Natural Resources

My team mined (type of pasta) _____

Number of Pasta Pieces Collected					
TRIAL 1	TRIAL 2	TRIAL 3	TRIAL 4	TRIAL 5	TRIAL 6

Create a bar graph to represent the number of pasta pieces your team mined in each trial.

NUMBER OF PIECES

TRIAL NUMBER

Chapter 9

Words to the Wild

By Karen Ansberry and Emily Morgan

Anotebook is perhaps the single most important piece of equipment a naturalist takes into the field. But notebooks are not only for use by field scientists: They are also excellent tools for helping students record observations outdoors, develop communication skills, and mirror the work of real scientists. They can also be a source of pride and delight, a personal and creative way for children to capture a moment of time spent in nature. They may contain observations and drawings of plants, animals, and their habitats; tallies, tables, and graphs; ideas and inferences; scientific questions and thoughtful "wonderings"; narratives and reflections; and even poetry. These activities offer some engaging ways for students to use notebooks to record both what they observe and what they think about nature.

Trade Books

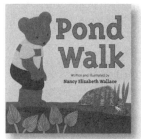

Pond Walk

By Nancy Elizabeth Wallace
Marshall Cavendish, 2011
ISBN 978-0-7614-5816-6
Grades K–4

SYNOPSIS
Buddy Bear and his mother spend a summer day observing and sketching pond life.

Salamander Rain: A Lake and Pond Journal

By Kristin Joy Pratt-Serafini
Dawn Publications, 2001
ISBN 978-1-58469-017-7
Grades K–4

SYNOPSIS
A boy named Klint joins the wetland patrol in which each member becomes an expert on a wetland habitat. This realistic, kid-friendly journal features paintings of lake and pond organisms, handwritten notes, clipped articles, and interesting facts about each creature. It provides a vibrant example that students can reference when creating their own nature notebooks.

Curricular Connections

Children develop understandings of life science concepts through direct experiences with living things and their habitats. Because kids today are spending less time outdoors, it is important for teachers to provide them with experiences that nurture their innate sense of wonder about nature. The *National Science Education Standards* (NRC 1996) suggest that nature study should include observations and interactions within the home, school, or local environment of the child. The Standards also point out that communication through speaking, writing, and drawing is a fundamental ability necessary to do scientific inquiry. The lessons presented here invite students to spend time in a local ecosystem recording observations, drawing pictures of what they observe, and communicating their thoughts and feelings about nature through writing. Check your school district's policies on taking students outside during the school day or away from the school grounds before doing the following activities.

Grades K–3: A Nature Walk

Materials

- Nature notebook (pp. 47–48; one per student)

Engage

Show students the cover of *Pond Walk* and introduce the author and illustrator, Nancy Elizabeth Wallace, who loves to go pond walking. Then ask students if they have ever been to a pond. What did they see, hear, touch, or smell? Have students share with partners. Read the entire book aloud. Tell students that Buddy is like a scientist: He makes observations using his senses, he shares his thoughts about what he observes, and he asks lots of questions. Ask, "What senses does Buddy use to make observations at the pond?" (sight, hearing, touch) and "How does he record these observations?" (drawing pictures and writing in his notebook). Point out examples of Buddy observing, thinking, and wondering, such as: *observing* the sound of geese honking; *thinking* that

when the water goes down, the lily pads do too; and *wondering* if there are any turtles in the pond. Tell students that like Buddy, they are going to have an opportunity to go to a natural place to observe, think, and wonder.

Explore/Explain

Choose a location on your school grounds, or plan a field trip to a local park, nature preserve, or wetland. Provide each student with a Nature Notebook and begin exploring the area together. Students should record the name of the natural place they are observing, the date, and the time. Then, have them sit quietly, choose a living thing from the area that they want to observe, and draw a detailed picture of it in the box on the cover of their notebook.

Next, explain that the inside of the notebook is divided into three parts, "What I Observe," "What I Think," and "What I Wonder." (See p. 47.) The left-hand page should contain only things that the students can directly observe with their senses. The right-hand page should contain thoughts and wonderings about those observations. For example, a student might observe a caterpillar crawling upside down on a leaf. After drawing a picture and recording observations of it, the student could write her thoughts about what she observes ("The caterpillar is really funny-looking! I think it can see me. It is moving fast. It might be scared.") and some wonderings ("I wonder how it crawls upside down? How many feet does it have?") Explain to students that it is important for scientists to know the difference between their observations (information gained through their senses) and what they think, feel, or wonder about their observations.

Elaborate/Evaluate

Writing a *syntu*—a type of poem usually written about something in nature—can help students reconstruct and reinterpret their notes into poetry form. A syntu has a simple five-line form with an emphasis on the senses. The first and last lines are one word; the other lines can be any number of words. On the back page of their nature notebooks,

Words to the Wild

have each student write a syntu following these guidelines:

Line 1: The name of a natural object, plant, animal, or place.

Line 2: An observation about line 1 using only one sense.

Line 3: A thought or feeling about line 1.

Line 4: Another observation about line 1 using a different sense than line 2.

Line 5: A synonym for line 1.

You can use the students' syntu poems to assess both the quality of their field observations and whether or not they understand the difference between an observation (Lines 2 and 4) and thoughts or feelings (Line 3).

Grades 4–6: Words to the Wild

Materials

- Spiral notebook (one per student)
- Samples of real scientists' field notebooks (pp. 49–51)

Engage

Introduce *Salamander Rain: A Lake and Pond Journal,* and ask "Have you ever kept a journal? What kinds of things would you write in a lake and pond journal?" Read aloud the "Official Planet Scout Journal" excerpt (p. 3), which introduces students to Klint and the purpose of his journal (to gather information about lakes and ponds to become an expert on these wetland habitats). Continue reading aloud several more pages from the journal. Have students point out the features that Klint has included, such as observations, measurements, and drawings of wetland animals; articles; interesting facts; and so on.

Explore/Explain

Tell students that like Klint, scientists use journals or notebooks regularly to record observations, questions, and ideas. Provide students with samples from

real scientists' field notebooks to read, like the ones on pages 49–51, then list and discuss some of their features (such as drawings, maps, tallies, weather data, narratives, and so on). Later, have students bring in notebooks and take them to a natural area on your school grounds, or plan a field trip to a local park, nature preserve, or wetland. Give students quiet time to explore and make some initial observations about the plants and wildlife they see. Next, have students choose one living thing on which to become an "expert" and write about in their own notebooks. Because scientists' explanations about what happens in nature come partly from what they observe and partly from what they think, it is important for students to be able to distinguish their observations from their ideas, feelings, and wonderings. One way to help them do this is to have students record only observations and drawings on the left-hand pages and anything else on the right-hand pages. When you return to the classroom, encourage students to do additional research on the organisms they observed and add that information to their notebooks.

Elaborate

Nature has been a source of inspiration to men and women throughout history. The following "words to the wild" are quotes from well-known people who spent their lives appreciating and studying various

"In every walk with nature one receives far more than he seeks." —*John Muir, naturalist and conservationist*

"The more clearly we can focus our attention on the wonders and realities of the world about us, the less taste we shall have for destruction." —*Rachel Carson, biologist, ecologist, and writer*

> "Conservation is a state of harmony between men and land." —*Aldo Leopold, ecologist and writer*
>
> "Only if we understand can we care. Only if we care will we help. Only if we help shall they be saved." —*Jane Goodall, naturalist and primatologist*

aspects of nature. Have students work in groups to restate the quotes in their own words and try to figure out their deeper meanings:

Next, have students choose their favorite nature quotes (these or others) to add to their notebooks or make up their own "words to the wild." Students can also do research to find out more about famous naturalists.

Evaluate

Have students share their notebooks with you, so you can assess whether or not they are able to distinguish between their observations and their ideas, feelings, and wonderings about nature. Observations (information gained through the senses) should be listed on the left-hand pages of their notebooks, and ideas, feelings, and wonderings should be on the right-hand pages. Also evaluate other aspects of the notebooks, such as quality and organization of observations, clarity of writing, or creativity.

Reference

National Research Council (NRC). 1996. *National Science Education Standards.* Washington, DC: National Academies Press.

Words to the Wild

WHAT I OBSERVE			
Sight	Sound	Smell	Touch

WHAT I THINK

WHAT I WONDER

My Nature Notebook

Name _____

Place _____

Date _____ Time _____

Drawing

My Nature Syntu

_____

_____

_____

_____

_____

Words to the Wild

JOURNALS 1 AND 2: Linette and Andy Sutphin, avian field biologists in Sheridan, Wyoming

Notes: Journal 1 is an aerial sage grouse lek survey with the names of each lek (e.g., West lek). A lek is a gathering of male sage grouse for the purpose of competitive display. Journal 2 is a mountain plover and raptor survey.

RTHA= red-tailed hawk
LEOW= long-eared owl
GOEA = golden eagle
ACTI = active nest

PON = ponderosa pine
JUN = juniper tree
CT = cottonwood tree

FEHA = feruginous hawk
GHOW = great-horned owl
MOPL = mountain plover

Journal 1

Sage grouse survey
4/26/07   Flight
Units VBARF, QC-9, Tipperary

0520 • West Lek   5 males displaying

0525 • Antelope Draw   GOEA over lek, 23 males displaying

0530 • Grub II   1 male displaying

0535 • Grub Draw 1 male displaying

0540 • Alvaro - inactive, no birds observed

0545 • Well lek - inactive

0550 • Dry Pond lek - inactive

Prairie Dog colony marking

Waypoint
016   - Active colony

617 - Active colony

018 - Active colony

019 - Active colony

Raptor nest surveys
20 · June 2007
unit · Iberlin

0640 - nest 1985, ACTI GOEA, imm. GOEA out of nest on cliff 200 m from nest tree.

0735 - Nest 2331, ACTI RTHA both adults challenging, 2 chicks in nest w/ some adult plumage ≈ 30 days old.

0810 · Nest 2335, ACTI LEOW, 4 owls flew from nest tree

0830 - nest 2340, GOEA nest empty, 2 adults flying in area w/in 200 yds of nest tree, no young observed, tree well leafed out.

0900 - nest 2336 ACTI LEOW, a tleast 4 owls flew from juniper tree

0955 nest 3267 ACTI GHOW, fledged no owls observed in area.

... 4 2 young

Journal 2

MOPL/ Raptor
CJA   5-27-07   Survey
Calm, p cloudy, 65° in am

06:30   Went up N495 Road in 51 76 Sec 05 (2-track)
Very steep draws; many JUN no PON or CT
Poor MOPL habitat

08:30
Wypt 009 424013 /4919204
Deep draw - along Cedar draw
T51 R76 Sec 02

Wypt 010 422857/4918940
51 76 02 ; 2 RTHA calling
Many possible nest sites

Wypt 011 0422680/4919367
Mid Survey reference pt.

Wypt 012 421326/ 4918859
End Survey pt.

CJA 5-27-07
Calm, p cloudy 80° @ 12:30

Wypt 013 423446/4919536
ACTI RTHA nest
PON   nest ht - 60ft
Sub ht - 80ft Aspect S
Ponderosa woodland
no young observed; adults calling

Wypt 014 424369/4919606
JUL/JUD Grassland
meadow 2 antelope
Sec 01   51 76

1430 - end Survey

©Susan C. Morse

March 10th '05

Yesterday's mid-March sun warmed up a few inches of fresh snow. By nightfall it was cold again and starry clear...perfect for bobcat courtship!

...This morning I tracked "Mystery", the tom, and his tracks showed where he abandoned his napping spot high on the cliffs... Along the way the sometimes hurrying tracks frequently stopped, backed up and paused beside rotten stumps, the undersides of logs or cliff walls. If you kneel down at such places and sniff the surface facing the hindfoot tracks you'll detect the distinctive odor of cat urine!

From the Field Notebook of Susan Morse
Naturalist, Habitat Specialist, and Founder of Keeping Track®
www.keepingtrack.org

©Susan C. Morse

August 10th, 2006

Day by day, year by year my data have given me the opportunity to transcend mere moments in time and comprehend whole patterns over many years, across vast expanses of habitat.

Today's bear mark trees add themselves to years of similar observations and cumulatively my data have taught us how to <u>predict</u> where to look for bear scent marking in the first place...a kind of search image. We're not just blundering through the forest bumping into these things! Thousands of these scent marking data entries have helped us appreciate that in northern New England black bears most often select to scratch, bite, and rub themselves on certain species of trees, and in certain specific situations within their habitat.

In this way we have learned that bears mark white birches along ridge line travel routes, balsam fir by the wetlands, and red pines along eskers. One simply has to look for these tree species within these specific habitats and "BINGO"- you will find evidence of scent marking bears. Half of tracking is knowing where to look and the other half is looking!

JOURNAL 3: Susan Morse, naturalist, habitat specialist, and founder of Keeping Track in Huntington, Vermont

Words to the Wild

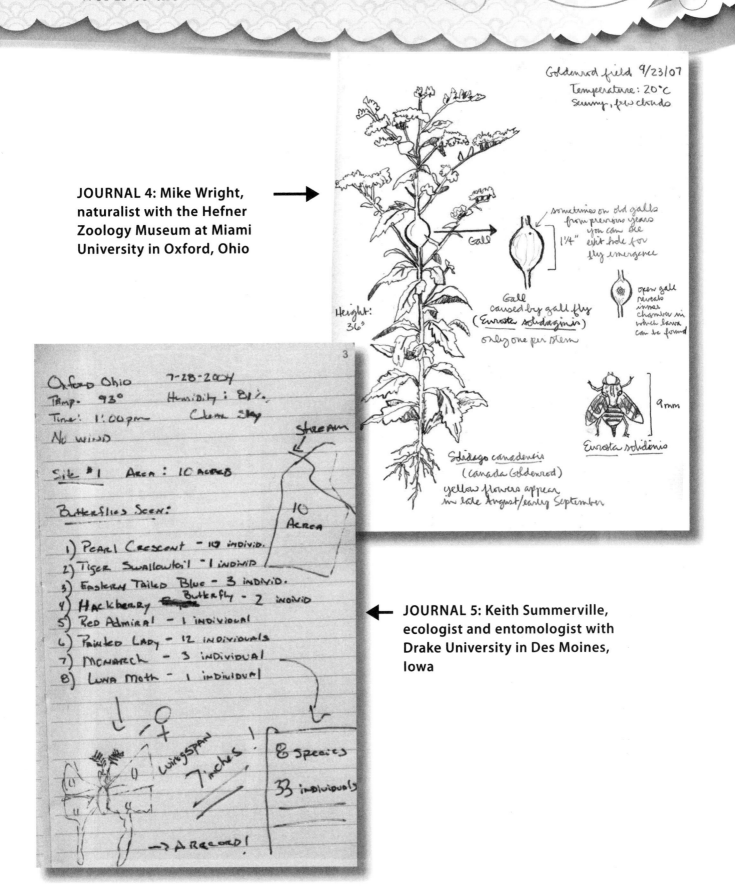

JOURNAL 4: Mike Wright, naturalist with the Hefner Zoology Museum at Miami University in Oxford, Ohio

Goldenrod field 9/23/07
Temperature: 20°C
sunny, few clouds

Gall

sometimes on old galls from previous years you can see 1¼" exit hole for fly emergence

Gall caused by gall fly (Eurosta solidaginis) only one per stem

open gall reveals inner chamber in which larva can be found

9mm

Eurosta solidaginis

Solidago canadensis (Canada Goldenrod) yellow flowers appear in late August/early September

Height: 36"

Oxford Ohio   7-28-2004
Temp. 93°   Humidity: 81%
Time: 1:00pm   Clear Sky
No wind

Site #1   Area: 10 acres

Butterflies Seen:

1) Pearl Crescent - 10 individ.
2) Tiger Swallowtail - 1 individ.
3) Eastern Tailed Blue - 3 individ.
4) Hackberry Butterfly - 2 individ.
5) Red Admiral - 1 individual
6) Painted Lady - 12 individuals
7) Monarch - 3 individual
8) Luna Moth - 1 individual

stream

10 Acres

wingspan 7 inches!

8 species
33 individuals

→ A Record!

JOURNAL 5: Keith Summerville, ecologist and entomologist with Drake University in Des Moines, Iowa

Chapter 10

Into the Woods

By Karen Ansberry and Emily Morgan

In today's electronic age of video games and MP3 players, children are not spending as much time outdoors as past generations did. Many fear that as a result, children are becoming increasingly alienated from the natural world. These trade book–inspired investigations encourage students to experience the excitement and mystery of their local ecosystems firsthand. In the process, they develop their skills of observation and learn to become better stewards of their environment.

Trade Books

In the Woods: Who's Been Here?

By Lindsay Barrett George
Greenwillow Books, 1998
ISBN 978-0-688-16163-7
Grades K–4

SYNOPSIS

Cammy and William don't see any wildlife as they walk through the woods, yet there are signs everywhere that animals have been around. They find such clues as an empty nest, a fallen branch with the bark gnawed off, and bleached bones. Each observation prompts the question, "Who's been here?" with the answer revealed on the following page.

The Woods Scientist

By Stephen R. Swinburne, photographs by Susan C. Morse
Houghton Mifflin, 2002
ISBN 978-0-618-04602-7
Grades 4–6

SYNOPSIS

This title from the excellent Scientists in the Field series features the work of Sue Morse, a forester, habitat ecologist, professional tracker, and passionate student of the woods. Readers experience the thrill of scientific discovery through her eyes and learn about the factors that led her to dedicate her life to wildlife conservation.

Curricular Connections

The *National Science Education Standards* (NSES; NRC 1996) suggest that students in grades K–4 learn to ask questions about objects, organisms, and events in the environment, particularly questions they can answer using reliable sources of scientific information and their own observations. In the primary grades, students should also develop such simple skills as observing and measuring to gather data and using magnifiers to extend the senses. In the K–3 lesson, students read the story and then explore "mystery objects" collected from the local ecosystem. Next, they explore a natural area outdoors, record their observations, and create a class book containing their own questions and answers about local wildlife.

In the 4–6 lesson, students learn about nature and conservation through the eyes of a real scientist. The *NSES* advise that students in grades 5–8 develop understanding of science as a human endeavor. Underlying this standard is the idea that women and men of various backgrounds, interests, talents, qualities, and motivations choose science as a career and devote their entire lives to studying it. Following this recommendation, students first read about the personal background and work of environmental ecologist Sue Morse, who has dedicated her life to conserving wild habitats. Next, they take a walk outdoors "in the shoes" of an environmental ecologist and then research local conservation efforts.

Check your district policy on taking students outside during the school day or away from the school grounds before doing the following outdoor activities.

Grades K–3: Who's Been Here?

Materials

- Mystery objects from the local ecosystem (e.g., plant galls, seed pods, owl pellets, and tree bark)

- Hand lenses
- Sticky notes
- Metric rulers
- My Mystery Object O-W-L chart (p. 57)
- 1 m pieces of string (one per student)
- Wildlife Questions student page (p. 58)
- Drawing paper

Engage

Introduce the book *In the Woods: Who's Been Here?* and build connections by talking about a time students may have walked in the woods or other natural area. Ask, "What wildlife or evidence of wildlife did you notice?" Explain that when you walk in the woods, you sometimes see wildlife, but more often you see *evidence,* or clues, that some animal has been there. Read the book aloud, prompting students to infer from the text and illustrations what animals had visited each area. Ask, "What can you observe in the picture? What clues can you get from the text? Which animals could have been here? Why do you think so?" Then reveal the picture of each animal.

Explore

Tell students that they are going to solve some mysteries just as the children in the book did. In advance, collect some "mystery objects" from your local ecosystem, such as plant galls, interesting seed pods, owl pellets, tree bark with bark-beetle tunnels, and so on. (If you cannot collect these items yourself, try borrowing some materials from a natural science professor or a museum.) Divide students into small groups and give each group a mystery object, a hand lens, a ruler, and an O-W-L (Observations, Wonderings, Learnings) chart (p. 57). Share some of your observations and wonderings about one of the objects: "This gall has a tiny hole in it, I wonder if something was living inside it?" Then ask students to draw, observe, measure, and discuss the objects with their groups as they fill in the first two columns of the chart. Invite students to share some observations and wonderings about their objects.

Into the Woods

Explain

Give each group some clues about its object—pictures, readings, or verbal hints. Have students write the information from their clues in the "Learnings" column of their chart. Invite each group to explain their inferences about the identity of their object and additional wonderings that were generated from the new information. Then reveal what the mystery objects are, where you collected the objects, and any additional information about them. Students can add this information to the "Learnings" column of the O-W-L chart.

Elaborate

Take students outdoors to observe nature closely (to a park or on a walk around the school grounds). Give each student a hand lens and a 1 m length of string to outline a circle on the ground. Explain that they will be working like scientists to draw pictures of what they see inside the circle and record their observations and wonderings. Then go outside and have students sit motionless for a while, listening and watching silently. Ask, "What do you hear, see, and smell? Do you see any evidence of wildlife?" Have them observe the living and nonliving things within their circle and make careful notes and drawings. Back in the classroom, have students share their observations and wonderings.

Evaluate

Discuss why questioning is very important in science. Questions help lead scientists to answers about the world. Scientists don't always find the answers to all of their questions, but they ask a lot of questions anyway. Help students create a "Question Book" about wildlife. In advance, collect a variety of nonfiction books and magazines on native wildlife. Invite students to read silently or in pairs, generating questions on sticky notes that they place on the pages of the book as they read. After reading time, have the class discuss some of the books and articles they read and their questions.

Next, pass out the Wildlife Questions student page (p. 58) and have each student choose a native animal they read about, make a detailed drawing of it, and write two or three interesting questions about it. Collect all of the student pages and bind or staple them together in a book. As students do more reading, they may discover answers to some of the questions. Encourage them to write the answers and sources on the backs of the book pages.

Grades 4–6: The Woods Scientist

Materials

- Mystery objects from the local ecosystem (e.g., plant galls, seed pods, owl pellets, and tree bark)
- Hand lenses
- Rulers
- My Mystery Object O-W-L chart (p. 57)

Engage

Have students discuss the following questions: "What do scientists do? Where do scientists work? What do scientists wear? What characteristics do you need to be a scientist?" (You may also want to have students "draw a scientist" to determine their perceptions of scientists before reading *The Woods Scientist*.) Tell students that you would like to share with them a book about a scientist who works outdoors, studies animals, wears hiking boots and comfortable clothes, and uses clues to solve mysteries.

Introduce *The Woods Scientist* and explain that the book is about Sue Morse, who is a forester, ecologist, professional tracker, and passionate student of the woods. Tell students that as they are reading, you would like them to jot down any traits or characteristics Sue has that make her well-suited to her job. (See pp. 4–9 about Sue's background and interests.) After reading, make a list of her characteristics on the board. These may include curious, athletic, loves learning and books, cares about

animals and nature, and so on. Discuss how these characteristics help Sue succeed in her profession, as well as which of these characteristics the students share. Next, tell students that Sue has dedicated her life to conserving wildlife. Ask them to listen while you read pages 10–13 for ways in which wildlife is being harmed by humans and ways in which Sue is helping wildlife. Finally, have students listen for ways Sue uses her scientific skills of observing and inferring to "read the forest" as you read about the black bear mystery (pp. 15–23) and the bobcat mystery (pp. 33–35).

Explore/Explain

Tell students that they are going to have the opportunity to solve some mysteries from the local ecosystem, just as Sue Morse solved the mysteries of the bear bite marks and the multiple bobcat tracks. (Do the "mystery objects" activity as discussed in Explore and Explain sections of the K–3 activity on the previous page.)

Elaborate

Next, tell students that they are going to "walk in the shoes" of a woods scientist by going outdoors to observe human impact on a natural place. Have students record observations in a journal as they answer questions such as the following: What wildlife can you observe? What clues or evidence do you see that wildlife has been there? What evidence of human impact can you see? How could this area be conserved? After the nature walk, have students research local environmental conservation efforts. The Nature Conservancy website (see Internet Resource) and your state's Department of Natural Resources website are both useful resources.

Evaluate

Refer back to the questions you asked students at the beginning of this lesson: What do scientists do? Where do scientists work? What do scientists wear? What characteristics do you need to be a scientist? Have students write about how their perceptions of scientists may have changed after reading about Sue Morse and participating in these activities. (You may also want to repeat the "draw a scientist" activity and have students analyze if or how their drawings may have changed.) Students can also create posters, brochures, or bulletin boards describing how people can help conserve local ecosystems.

Reference

National Research Council (NRC). 1996. *National science education standards*. Washington DC: National Academies Press.

Internet Resource

Nature Conservancy
    www.nature.org/wherewework/northamerica/ states

Name: _____

My Mystery Object

Drawing

O	W	L
What do you **OBSERVE** about the object?	What do you **WONDER** about the object?	What did you **LEARN** about the object?

Name: _____

Wildlife Questions

Name of Animal: _____

My questions about this animal:

_____

_____

_____

_____

_____

_____

_____

```
Drawing




```

Chapter 11

Discover Reading

By Christine Anne Royce

We often gloss over the history of science—the women and men who have made advancements in the area of scientific discovery. These notable individuals are the backbone of our field. In these activities, we honor these scientists by encouraging children to read about their life stories and make their own discoveries.

Trade Books

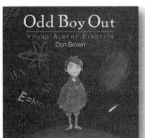

Odd Boy Out: Young Albert Einstein
By Don Brown
Houghton Mifflin Books for Children, 2004
ISBN 978-0-618-49298-5
Grades K–6

SYNOPSIS
Don Brown chronicles Albert Einstein's life from birth to adulthood in a creative way that describes the scientist's personality to the reader, explaining how he was not a great student and even disliked school. The author reveals that Einstein was supported by his parents even though he was considered, as the title suggests, an "odd boy."

Reaching for the Moon
By Buzz Aldrin, paintings by Wendell Minor
HarperCollins, 2005
ISBN 978-0-06-055445-3
Grades 1–5

SYNOPSIS
Buzz Aldrin, space pioneer on the first lunar landing, recounts specific episodes throughout his life that influenced him to pursue a career as an astronaut. Anecdotes include Aldrin getting his nickname, riding in an airplane for the first time, and eventually joining the Gemini and Apollo programs. The biography highlights how many small events played a role in helping Aldrin achieve a larger goal.

Curricular Connections

Researchers say students need to be able to read for content (i.e., efferent reading) and to experience, think, and feel while doing so (i.e., aesthetic reading; Rosenblatt 1991). The *National Science Education Standards* state that "students need to understand that science reflects history and is an ongoing, changing enterprise" (NRC 1996, p. 107). The Standards also infer that by teaching students about scientific ideas of the past, we lay the foundation for the ideas they may develop in later years.

Odd Boy Out and *Reaching for the Moon* teach younger students that "science and technology have been practiced by people for a long time" and that "many people choose science as a career and devote their entire lives to studying it." (NRC 1996, p. 141) In addition, children can relate to these stories because they focus on the lives of both scientists as youths. See Figure 11.1 for more recommended books. For additional ideas for researching female scientists, see Campbell (2007).

Regardless of which scientist a student investigates, these activities provide an opportunity for students to learn more about individuals who contributed to science and perhaps start to see themselves as having similar experiences—and potential.

Grades K–3: Biography Boxes

Purpose

Students will investigate the life of a scientist and create a timeline that outlines the key events in this person's life.

Materials

- Biographies or autobiographies of scientists
- Sentence strips
- Cardboard or plastic storage boxes large enough to hold the book and sentence strips for each student

Figure 11.1. More Recommended Reading

Adler, D. 1996. *A picture book of Thomas Alva Edison.* New York: Holiday House.

Brown, D. 1999. *Rare treasure: Mary Anning and her remarkable discoveries.* Boston: Houghton Mifflin.

Grimes, N. 2002. *Talkin' about Bessie: The story of aviator Elizabeth Coleman.* New York: Orchard Books.

Johnson, D. 2006. *Onward: A photobiography of African American polar explorer Matthew Henson.* Des Moines, IA: National Geographic Children's Books.

Krull, K. 2007. *Giants of science: Marie Curie.* New York: Penguin.

Lasky, K. 1995. *She's wearing a dead bird on her head!* New York: Hyperion.

Martin, J. B. 1998. *Snowflake Bentley.* New York: Houghton Mifflin.

Sis, P. 1996. *Starry messenger.* New York: Frances Foster Books.

Wadsworth, G. 2003. *Benjamin Banneker: Pioneering scientist.* Minneapolis, MN: Lerner Publishing Group.

Yolen, J. 2003. *My brothers' flying machine: Wilbur, Orville and me.* New York: Little Brown.

Procedure

1. Ask the students to explain the difference between *biography* and *autobiography* to begin your discussion. Once they have provided definitions for these two terms, ask them to name scientists they have heard of who might have a biography or autobiography written about them.

2. Read aloud one or both of the selected books, asking students to listen to the type of information that is presented in the book (e.g., places, dates, what the scientist did as a child, if they had any siblings). Discuss what type of information helps a reader know about a person's life.

Discover Reading

3. As a class, generate a list of scientists or scientific discoveries students might want to learn more about. Possible topics might include: "Who figured out what snowflakes look like?" and "Who made the first flight and how did they do it?"

4. Once students as individuals, pairs, or groups have selected scientists and located appropriate biographies, provide them with sentence strips and ask them to record in their own words seven to nine key events in each scientist's life on a separate sentence strip. Students should also locate or draw a picture of their scientists.

5. After each individual or group has completed their research, the teacher can assemble "biography boxes" with the picture and name of the scientist and sentence strips (preferably laminated after completion). If available, the book that was used can be placed in a learning center for future reading time. These individual biography boxes can then serve as learning centers for other students.

6. After students have viewed each biography box, talk about what types of things the different scientists did to develop careers in science, such as "How did they study the topic they were interested in?"

7. Once students have further explored the lives of the scientists, ask them to think about areas of science they like, for example, the solar system, dinosaurs, insects, or simple machines. Ask them what they could do to learn more about these topics and what types of activities they could do (with parental permission of course) to begin to learn more about their interests.

8. As students explore science concepts throughout the year, they can list their own accomplishments in a similar format and begin to identify themselves as scientists.

Figure 11.2. Albert Einstein Biography Box (using information from Odd Boy Out)

He liked math in school as a young boy.

Max Talmud gives Albert a geometry book.

Albert Einstein won the Nobel Prize for his work.

Grades 4–6: Learning About a Life

Purpose
Students will research a scientist's life and share information with their classmates.

Materials
- Internet access
- Chart paper or poster board
- Markers
- Biographies or autobiographies about scientists

Procedure

1. Read aloud one or both of the selected books, asking students to seek the following types of information: where the scientist lived, what interested him in science, and key events that helped him in his career.

2. Have students list their information collectively on chart paper or individually in a science notebook. Next, have them categorize the information into topics such as "What helped the scientist become interested in the topic?" and "Information about the scientist's life." An example for the first category could be when Buzz's father gives him a compass. This activity helps students begin to understand the different types of information that can be obtained from nonfiction.

3. Reread the book to the students, and ask them not to focus on the facts as they did the first time, but rather to think about how they feel about the events and the person. An example in *Reaching for the Moon* is when Aldrin discusses being taken flying for the first time at the age of two and being frightened but also thrilled. This is similar to a text-to-self approach in a read-aloud in which students try to connect what the character is experiencing to their life. This allows the students to be engaged in an example that is an aesthetic form of reading.

4. Ask students to identify parts of the book in which they could connect or relate to the experiences that the scientist had and explain how they felt and why.

5. Next, have each student select a scientist (you may want to provide a list based on your curriculum) and investigate that person's life using the internet and books. They should develop a creative way to present the information to the class (e.g., a display or skit). The final product should contain key facts about the scientist's life, his or her contribution to science, what interested him or her in science, and the student's reason for choosing this scientist to research.

6. This project will most likely take several days and several drafts, allowing the teacher to check on the students' progress in small groups and assess understanding of reading from an efferent or aesthetic approach.

7. Once the students' research has been completed and their projects developed, schedule time for them to present their findings to the rest of the class in a book discussion. It may be helpful to have the students develop one or two key questions ahead of time for their classmates to focus on during their presentation.

References

Campbell, A. 2007. Weaving women into the curriculum. *Science Scope* 31 (2): 54–58.

National Research Council (NRC). 1996. *National science education standards*. Washington, DC: National Academy Press.

Rosenblatt, L. M. 1991. Literature—S.O.S.! *Language Arts.* 68 (6): 444–448.

Chapter 12

How It's Made

By Karen Ansberry and Emily Morgan

In colonial times, children knew a lot about the products they used in their daily lives. There were no factories for making things, so their families and communities made furniture, wove cloth, and sewed clothes. Children made toys from natural objects or bits of string and cloth. Instead of pencils, students used homemade quill pens created from goose feathers. Your students are most likely not knowledgeable about the raw materials, the design processes, and the technology involved in manufacturing the products they use every day. Making connections with local manufacturers can connect students to the real science and engineering happening in their communities. Both of the lessons presented here suggest inviting local engineers and other manufacturing professionals to share how they use science and technology to design and make products that solve human problems and enhance the quality of life.

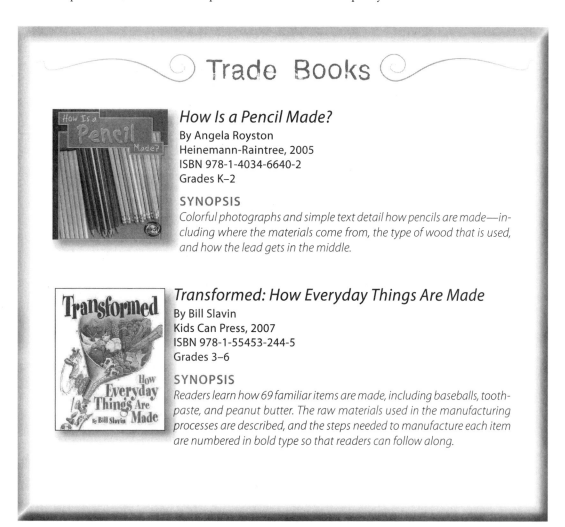

Trade Books

How Is a Pencil Made?
By Angela Royston
Heinemann-Raintree, 2005
ISBN 978-1-4034-6640-2
Grades K–2

SYNOPSIS
Colorful photographs and simple text detail how pencils are made—including where the materials come from, the type of wood that is used, and how the lead gets in the middle.

Transformed: How Everyday Things Are Made
By Bill Slavin
Kids Can Press, 2007
ISBN 978-1-55453-244-5
Grades 3–6

SYNOPSIS
Readers learn how 69 familiar items are made, including baseballs, toothpaste, and peanut butter. The raw materials used in the manufacturing processes are described, and the steps needed to manufacture each item are numbered in bold type so that readers can follow along.

Curricular Connections

The *National Science Education Standards* (NSES) suggest that students in grades K–4 should develop the ability to distinguish between the natural world and the designed world. In the K–2 lesson, students sort objects as natural (occurring in nature) and designed (made by people). In the lesson for grades 3–6, students learn how pencils and a variety of other objects are made from raw materials. The *NSES* also suggest that students understand that people are continually inventing new products and new ways of doing things and that many people find science and engineering to be enjoyable careers. Bringing local engineers and manufacturing professionals into the classroom can help your students better understand design and manufacturing processes and the variety of people and jobs that are involved in making products.

Grades K–2: From Tree to Pencil

Materials

- Log
- Pencil
- Bag
- Natural objects (e.g., feather, rock, tree branch, apple, leaf, seashell, salt crystals, sand, and water)
- Designed objects (e.g., CD, crayon, birthday candle, pencil, cotton cloth, and plastic toy animal)
- Poster paper
- Markers or crayons

Engage

Show students a log and a pencil. Ask them what the objects have in common (i.e., they are both made of wood). Tell them that making pencils begins with a piece of wood. Ask, "How do you think a piece of wood becomes a pencil?" Have them turn and talk. Then ask, "What other things

are part of a pencil besides wood?" (metal, rubber eraser, paint, lead). "How do you think the lead gets inside the pencil?" Tell students that there are two types of objects or materials, those that are natural (like the log) and those that are designed (like the pencil). Explain that *natural* refers to not only living things like trees and animals, but also materials found in the Earth, such as rocks, metals, and chemicals. *Designed* refers to things that have been made by people to solve a problem or make life better. Have students look around the room and name things that people designed. They will realize that this includes almost everything in the room. Challenge them to find something in the room that was not designed or made, such as a classroom pet, plant, apple, or rock.

Explore

Give students a bag containing both natural and designed objects to sort through (or pictures of them). Natural objects may include a feather, rock, tree branch, apple, leaf, seashell, salt crystals, sand, and water. Designed objects may include things such as a CD, crayon, birthday candle, pencil, cotton cloth, or plastic toy animal. Have students sort the objects (natural or designed) and then justify how they sorted. Next, tell students that all designed objects start out as natural materials, for example, a pencil starts out as a piece of wood. Have them determine what natural materials each designed object is made from. Tell students that they are going to learn the steps involved in turning a natural material (wood) into a designed product (pencil).

Explain

Show students the book *How Is a Pencil Made?* Have students describe some of the features that indicate it is a nonfiction book—table of contents, bold-print words, index, and glossary. Model how to use the table of contents in a nonfiction book to locate specific information. Tell students that you will read aloud the chapters that describe how a pencil is made. Have students describe the steps involved in making a pencil as you create a flowchart on the

How It's Made

board (e.g., cutting down incense cedar trees is the first step and attaching the eraser with a metal band is the last step).

Elaborate

Invite a local engineer or manufacturing professional to share with your class how the products he or she designs or manufactures are made. Ask your guest to include the following in the presentation: a sample or photograph of the product he or she designs or makes, the natural materials the product is made of, how the product was designed by people, how it solves a problem or enhances the quality of life, the steps involved in making the product, how his or her job is part of creating the product, and the qualities that make a person successful in engineering or manufacturing. If possible, arrange a follow-up field trip to the company or factory, so students can observe the manufacturing process firsthand (for virtual fieldtrips, see Internet Resources).

Evaluate

As a class or in teams, have students create a poster that describes how the product shared by the guest is made. The poster should include a list of the natural materials, the steps, and the different jobs of the people involved in making the product.

Grades 3–6: Everyday Things

Materials

- Some products featured in *Transformed: How Everyday Things Are Made* (e.g., marble, plastic dinosaur, wool cloth, cotton cloth, crayon, pencil, plastic wrap, adhesive bandage, and pasta)
- Poster paper
- Markers or crayons
- Resources to research how products are manufactured

Engage

Ask, "What is the world's only portable, lightweight invention that can draw a line 35 miles long, write an average of 45,000 words, absorb 17 sharpenings, and delete its own errors?" The answer is, of course, a pencil! Then ask, "How do you think a pencil is made?" Give students time to turn and talk to a partner and then share their ideas with the whole group. Next, introduce the book *Transformed: How Everyday Things Are Made* and read pages 62 and 63 about how a pencil is made. You may also want to share the "Let's Make a Pencil" video (see Internet Resources) showing pencils being made in a real factory.

Explore/Explain

Tell students that they are going to learn how some other objects are made. Give each group of 3–4 students some of the products featured in the book *Transformed: How Everyday Things Are Made*, such as a marble, plastic dinosaur, wool cloth, cotton cloth, crayon, pencil, plastic wrap, adhesive bandage, or pasta. Allow time for students to observe the items and explain how they think each one might have been manufactured. Then, read the first paragraph in the book about each product (leaving out the name of the product) and have students try to figure out which one it is. Continue reading the steps of the manufacturing process until all students can infer which product is being described. Have students explain their thinking.

Elaborate

Find out what materials or products are designed or manufactured in your community. Invite an engineer or other manufacturing professional from a local company to share with your class how the products he or she designs or manufactures are made. Ask your guest to include the following in the presentation: a sample or photograph of the product he or she designs or makes, the natural materials the product is made of, how the product was designed by people, how it solves a

problem or enhances the quality of life, the steps involved in making the product, how his or her job is part of creating the product, and the qualities that make a person successful in engineering or manufacturing. If possible, arrange a follow-up field trip to the company or factory, so students can observe the manufacturing process firsthand (for virtual fieldtrips, see Internet Resources).

Evaluate

Have each student choose an everyday object, research how it is made, and then create a poster (similar to the two-page spreads in the book *Transformed: How Everyday Things Are Made*). Their posters should include the materials used to create the object, the steps in the manufacturing process, drawings or photos of the process, and the different jobs that are involved in creating the product.

Reference

National Research Council (NRC). 1996. *National science education standards.* Washington, DC: National Academies Press.

Internet Resources

How Everyday Things Are Made
  http://manufacturing.stanford.edu

How It's Made Videos From Science Channel (preceded by a brief commercial)
  http://science.discovery.com/videos/how-its-made

How Products Are Made
  www.madehow.com

Let's Make a Pencil
  www.pencils.com/lets-make-pencil-video

Chapter 13

It's About Time

By Christine Anne Royce

"Is it time yet?" is a cry heard from many a student's mouth. Is it time for lunch? Is it time to go home? Is it time for recess? Is it time for science? (We can always hope.) The list can go on and on. Students often equate time with an action or event at a particular point; however, time is much more than that—it is the concept by which we track everything in history. In these activities, the focus is on helping students understand methods by which we track and tell time.

Trade Books

How Do You Know What Time It Is?
By Robert E. Wells
Albert Whitman and Company, 2002
ISBN 978-0-8075-7940-4
Grades K–3

SYNOPSIS
The author provides an overview of how we have measured time throughout the ages up through today—from shadow sticks to atomic clocks. The illustrations describe what can be an abstract concept for young students and provide examples and analogies of how time is important to their lives.

On Time: From Seasons to Split Seconds
By Gloria Skurzynski
National Geographic Children's Books, 2000
ISBN 978-0-7922-7503-9
Grades 4–6

SYNOPSIS
This book examines the way mankind has told time throughout the ages. The reader is provided with an overview of the different types of time systems, including calendars, seasons, and different types of clock systems. Skurzynski also provides depth, helping readers understand the historical connections to important scientists. This book is easy to read and the illustrations aid understanding of the concept.

Curricular Connections

Time is one of those elusive concepts that students know exist, but they do not necessarily understand what the passage of time means. Nor do they understand how time is measured. Students often ask, "Is it time yet?" or, "How much time do we have?" The answers to those questions do not help students connect the idea that there are different ways of counting time or measuring the passage of time.

The *National Science Education Standards* do not specifically mention time as a concept that needs to be covered in the content standards. However, the concept of time is related to many of the other content standards, including Science as a Human Endeavor, which discusses how "science and technology have been practiced by people for a very long time" (NRC 1996, p. 141), and the Earth and Space Science standard that states "objects in the sky have patterns of movements" (p. 134). These types of concepts do not at first seem connected to the idea of time, but students use concepts in time to fully understand the content standards. For example, the fact that objects in the sky have patterns of movement is closely connected to seasons, including historical methods by which time was calculated using the lunar phases and the idea of a year, which was defined as the Sun returning to the same point in the sky being a unit of time.

Regardless of which unit of time is used to help students understand that time passes, it all connects back to the idea that humans use newer and better improved methods by which to tell time. Once upon a time, time was measured rather "roughly" resulting in time being off and needing to be "adjusted" at varying points in our past. Today, time is measured precisely using inventions such as the atomic clock.

Grades K–3: My Time

Purpose

To help students begin to understand different units of time

Materials

- Chart paper
- Markers
- Clock with second hand
- Paper calendar

Procedure

1. Begin by asking students what *time* is and recording their answers on the board. One of the more common answers is "when we are supposed to do something." While they are providing their ideas, ask how they "measure" time. Students most likely will say with concrete objects such as clocks or watches or cell phones.

2. Once their initial ideas are recorded, do a walk-through of the book *How Do You Know What Time It Is?* and ask students to refer back to their list of ideas. Then conduct a read-aloud of the book focusing on the following points: On page 5, have students speculate as to what might have been used before we had clocks. On pages 20–21, Ask the students if they track any types of time on a calendar or through day and night. Birthdays are a common idea that crops up when they begin to think of a calendar as an instrument that measures time.

3. Once the book has been read, ask the students to revisit their list of ideas and questions. Can they add any new answers to "What is time?" and "How do we measure time?" At this point, they will most likely still stick with the concrete ideas of a calendar, the seasons, or a sunrise or day. Introduce the idea that the students are going to look at different ways to measure the same amount of time.

4. Begin by having students focus on the clock (a clock with a second hand is needed) and point out the second hand, explaining that each "click" of the second hand is a unit of time we call a second. Even though some students may already know that there are 60 seconds in a minute, ask

them to count each click until the hand makes one complete cycle around the clock (thus resulting in 60 clicks or one minute). This can be done aloud to help students keep track of their counting.

5. Next, ask the students to predict or state how many minutes are in an hour (the next largest unit of common time). They can arrive at this by counting the number of times the second hand sweeps past between two of the larger numbers (which is five times) and then use simple addition or multiplication to determine that there are 60 minutes in one hour. Continue to have discussions about the next largest unit of time—seconds to minutes, minutes to hours, hours to days, days to weeks, weeks to months (which can be very tricky since the number of days, and thus the number of weeks, in a month is not constant) or weeks to years.

6. Using concrete items that will allow students to count or verify their predictions will help them—for example cheap paper calendars will allow them to count the number of weeks in a year. While using the calendar, students can record their birthdays and ages and discuss who is older—a concept of time for age.

7. After the discussion, ask the students to pick one idea of time, and then with classroom aides or parental support, create an illustration that shows that unit of time and how old they are in that unit of time. For example, rather than simply saying "seven years old," they could say "seven times 52 weeks old." This will incite the class to discuss things like leap year or being seven years old and 26 weeks rather than just seven and a half.

This topic can be a bit of a brain teaser for children depending on their age and cognitive development. As the classroom teacher, you know your students best and should only go as far as they are able to go at this time!

Grades 4–6: Build a Clock

Purpose

To experiment with pendulums as a measure of time

Materials

- Lead weights or sinkers (teardrop-shaped weights often called "bank sinkers" in fishing equipment)
- String
- Table
- Books
- Stopwatch
- Measuring tape
- Building a Clock student page (p. 71)

Procedure

1. After students read the book *On Time: From Seasons to Split Seconds*, ask them to recall the section on mechanical clocks and Galileo's use of the pendulum as a method of measuring time. Ask, "How is a mechanical clock different from other methods of measuring time (such as seasons or lunar cycles)?" Students often respond that this is a "manmade" method or that it's more accurate than other methods.

2. Explain to the students that they are going to perform the same kind of investigations with pendulums as Galileo, and later Christiaan Huygens, did in trying to devise a pendulum that has a period of one second.

3. Provide the students with weights (lead fishing weights work); an ample supply of string, scissors, and measuring tapes; and a watch with a second hand or a stopwatch. Each group of two to four students will need an area to conduct their investigations, such as a table. Ask them to create a data table or give them the Building a Clock student page. Have them choose three different lengths of string (with guidance) for

their pendulum. (The length of a pendulum with a period of one second is about 39 inches.)

4. Students should place the bob (lead weight) on the end of the pendulum (string) and hang it over the side of a table. They can use books to hold it in place so that it doesn't slip or gain length. One student pulls back the pendulum about 12 inches from its stationary position and releases it so that it will start to swing. Another student counts the number of swings back and forth. A third student—or the person who started the pendulum swinging if students are working in pairs—times the pendulum for 60 seconds. Students record the number of swings in 60 seconds, repeat twice more for this release height, and find an average.

5. To help students confront the idea that only the length of the pendulum affects its period, ask them to repeat step 4, but have them pull the pendulum back only six inches, which results in the pendulum swinging through a smaller arc (or not going as "high" in student language).

6. Repeat step 4 again for the additional two lengths chosen for the experiment. Ask the students what happens as the string gets shorter (it will swing faster in the same amount of time)

and as the string gets longer (it will swing slower in the same amount of time).

7. Now that students have some data points, ask them to predict how they can adjust the length of their pendulum to obtain 60 swings in 60 seconds, which means that each swing would be 1 second. Some adjustment of the length will be necessary.

8. Return to a full-class discussion to report the results to the class and compare. Ask the students if they would have liked to have this type of instrument to tell time all day. (Note to teacher: There are other forces that will eventually slow down their pendulum clocks. The same forces affect clocks that use pendulums or pendulums and gears.)

Reference

National Research Council (NRC). 1996. *National science education standards*. Washington, DC: National Academies Press.

Internet Resource

How Pendulum Clocks Work
  http://electronics.howstuffworks.com/clock.htm/ printable

It's About Time

Building a Clock

	TRIAL #1: LENGTH = ___ INCHES	TRIAL #2: LENGTH = ___ INCHES	TRIAL #3: LENGTH = ___ INCHES
Number of swings that occurred in 60 seconds	Attempt #1 Attempt #2 Attempt #3		
Average number of swings after three attempts			
Modification to pendulum	Based on our results, our group is going to adjust the length of the string by lengthening it shortening it because _____ _____ _____ _____ _____	Based on our results, our group is going to adjust the length of the string by lengthening it shortening it because _____ _____ _____ _____ _____	Based on our results, our group is going to adjust the length of the string by lengthening it shortening it because _____ _____ _____ _____ _____

Chapter 14

If You Build It ...

By Christine Anne Royce

From the youngest ages, children construct buildings, bridges, towers, and anything else that comes to mind using a variety of materials. These books and activities take this interest in construction and build on it by allowing students to experiment with structures, their design, and how they withstand forces. Through building activities young students start to develop understanding of the science concepts associated with engineering and technology, while older students are able to put their knowledge of structures to the test in the design process.

Trade Books

How a House Is Built

By Gail Gibbons
Holiday House, 1996
ISBN 978-0-8234-1232-7
Grades K–3

SYNOPSIS

Readers are taken on a journey as a house is built from the ground up. Through examining the different parts of a house and the work of contractors and architects, the reader is able to determine the important aspects of building a structure. Very colorful illustrations and captions help the reader develop understanding.

Bridges! Amazing Structures to Design, Build, and Test

By Carol A. Johmann and Elizabeth J. Rieth
Williamson Publishing, 1999
ISBN 978-1-885593-30-6
Grades 4–6

SYNOPSIS

This book combines the subjects of science, history, and technology while presenting activities that help students examine the design and construction of bridges. It introduces key vocabulary and provides both real pictures and illustrations to help convey the concepts. This book provides a large amount of information and references on the topic of bridges.

Curricular Connections

The idea of building structures is one that starts early in a child's development through play with blocks. *How a House Is Built* and *Bridges! Amazing Structures to Design, Build, and Test* will allow students to use creativity and curiosity to engage in the development of different types of structures as they test their designs against natural forces that a house may need to withstand.

This intrigue with building allows students to investigate the intersection of Science and Technology as described in the *National Science Education Standards* (*NSES*) where students "establish connections between the natural and designed worlds [and are provided] with opportunities to develop decision-making abilities" (NRC 1996, p. 106). The *NSES* are quick to point out that at this intersection, "science as inquiry is parallel to technology as design. Both standards emphasize student development of abilities and understanding" (NRC 1996, p. 107). Students will make decisions about the best design of their structures through the trial-and-error process—a large part of inquiry. By allowing students to use common, familiar objects such as building materials, they are put in a safe environment that allows them to explore.

How a House Is Built presents the opportunity for young children to examine parts of a house and then try to build a structure based on the information presented. The investigating of how to build a structure is part of the design process. As children get older, they are better able to examine the design process, to think about what they already know and apply it to their designs before they build and test it, as described in *Bridges! Amazing Structures to Design, Build, and Test.*

Grades K–3: Let's Build a House

Purpose

To experiment with different structures and forces through play

Materials

- Wooden building blocks or other stackable blocks of varying size (6–10 per group)
- Straws
- Paper
- Clay
- Corrugated cardboard
- Tape
- Plastic blocks
- Hair dryer

Procedure

1. Before introducing *How a House Is Built*, have students find pictures of houses, towers, bridges, and other structures in magazines or on the internet and create a large mural of these structures.

   Ask students to generate a list of ideas about how the different structures are built. Ask, "What do you notice about the base or bottom of the structure? What are the different shapes you see in the structure?" The construction process relies heavily on the use of forces and balance, and a sturdy foundation for the building is crucial.

2. Allow the students to examine the different parts of a house and explain what each part is for in the house design. For example, the roof rafters support the actual roof that is placed on the house. Challenge the students to think about what kind of different events a house must withstand to survive. Possible answers include tornadoes, hurricanes, and earthquakes. Without scaring young children, explain that houses are built to withstand these different forces, but the design of the house is important to that survival.

3. Once students have had a chance to consider how a house is built, challenge them to design houses based on what they know. They can begin by sketching out their houses on paper and then building them with the supplies provided. Some

If You Build It ...

students may wish to build houses of blocks while others might want to use cardboard. (Enlisting parent volunteers to assist with the actual "construction" may be helpful, as students should be carefully watched while working with scissors.) While the students are building their houses, ask them to consider what they are doing to help the house survive a windstorm or earthquake. If necessary, ask students to return to the pictures they found and see if they have a match for a structure. Ask the students to speculate about why this structure design is sturdy. Ideas may include, "There is a larger base than top" and "The buildings are more square or rectangular rather than circular."

4. Once the students have constructed their houses, put their sturdiness to the test. The first test will be a tornado (i.e., hair dryer). Have students put their houses on a desk in the front of the room and check for sturdiness against "wind" by turning the hair dryer on each house. The teacher should keep the hair dryer on the same setting and aim it at the same place for each student's house. See what "damage" is done to the house after one minute and then two minutes. Have the students make observations about where the damage was, if there was any.

5. After the wind test, the houses can be subjected to an earthquake test. Again, have students put their houses on a desk. The teacher should shake the desk with a fair but not extreme amount of force. After one shake, check to see what damage occurred. If the houses survived, shake the desk for 15 seconds and then check what damage occurred.

6. Students can return to the illustrations in the book to determine how houses are made sturdy to withstand different types of natural disasters.

Grades 4–6: Build a Bridge

Purpose

To construct a bridge using common materials and determine how much of a load it will hold

Materials *(per group of two to three students)*

- 26 pieces of white copy paper (20 for the initial planning activity / six for the final bridge)
- Paper for recording ideas and designs
- Roll of tape
- Scissors
- 50–100 pennies

Procedure

1. Pose the following challenge to the students: "Can you design a bridge that will cover a 22.5 cm (9 in.) span that will support a load using only copy paper?" Break students into teams of two or three and allow them to find a place to work.

2. Provide each team with supplies and explain that these materials are only for the planning part of the project. They will be given additional paper to construct their actual bridges at a later point. Explain that students must construct bridges to span 22.5 cm using ONLY six sheets of paper. They may use tape to assemble the bridge but may not secure the bridge to any surface. Each bridge span will be suspended using two stacks of textbooks of the same height. The goal is for the bridge to hold as many pennies as possible.

3. Students can use *Bridges! Amazing Structures to Design, Build, and Test* to examine illustrations and individual aspects of bridge design and to learn vocabulary.

   The students should make, test, and explain the reasoning behind at least three bridge designs. They should record their ideas (the design phase) before they actually assemble the

bridges; draw their bridges and keep the sketches for demonstration purposes (the build phase); and record their data for how much weight the bridges held as well as observations they made about where the bridge failed (the test phrase).

4. Finally, after allowing students to experiment with their designs, have the students choose their final designs and build the bridges using the specifications outlined previously. When all groups have constructed their bridges, have each group describe its trials, explain what it learned from each attempt, and present its final design. Then have a competition to determine which bridge design can hold the largest load of pennies.

5. After a bridge winner has been determined, ask the students to discuss what they learned from the different designs—focus on construction techniques, shapes, strength of the bridge, and so on. For example, when the paper is folded in certain ways, it makes the bridge stronger; also, the bridge needs to be supported somehow or it will sag in the middle. A bridge building competition always excites students and engages them in the design process.

A familiar movie phrase states, "If you build it, they will come." A slightly modified phrase fits the purpose and use of these activities in the classroom—"Students will come to understand important science concepts, if you let them build."

Reference

National Research Council (NRC). 1996. *National science education standards.* Washington, DC: National Academies Press.

Chapter 15

Imaginative Inventions

By Christine Anne Royce

*I*nvention assists students in understanding the relationships between the individual subjects of STEM (science, technology, engineering, and mathematics) education. The playful nature of these trade books adds that additional spark of creativity that is needed in the invention process.

Trade Books

Imaginative Inventions
By Charise Mericle Harper
Little, Brown Books for Young Readers, 2001
ISBN 978-0-316-34725-9
Grades K–4

SYNOPSIS

This colorfully illustrated book will engage young children who want to know where common things came from. In all, 14 different inventions—ranging from potato chips to roller skates to the vacuum cleaner—each occupy a two-page spread in this book.

So You Want to Be an Inventor?
By Judith St. George, illustrated by David Small
Puffin Books, 2005
ISBN 978-0-14-240460-7
Grades 2–6

SYNOPSIS

The reader is introduced to inventors and inventions, from automobiles to lightbulbs to Velcro, in this whimsical and fun narrative. Using creative illustrations, the book suggests, "If you want to be an inventor, find a need and fill it." The author then provides examples of inventors who took that advice and shared what they developed.

Curricular Connections

Inventors think of how to improve some aspect of everyday life, then build a device or design a process to fill that need. Each of these concepts is related to the design process. Grades K–3 can begin to (1) increase their understanding of the design process and make devices to answer a specific purpose; (2) compare and contrast two objects that are used for the same purpose; or (3) construct simple structures to solve a problem. *Imaginative Inventions* walks the reader through such examples (e.g., wheelbarrow or eyeglasses).

In *So You Want to Be an Inventor?* more elaborate inventions are described, ranging from dynamite to computers. What are not presented are the trials that happened prior to the finished product, pointing out that "perfectly designed solutions do not exist" and that "all technological solutions have trade-offs" (NRC 1996, p. 166). In the activity for grades 4–6, students will engage in a process in which they must consider these trade-offs as they build their device.

Grades K–3: What Was It Used For?

Purpose

Students will begin to think about the purpose for different objects that they may or may not recognize.

Materials

- A collection of different objects found around the house or in the garage that students do not immediately recognize (e.g., a handheld egg beater, egg separator, lobster measurer, orange peeler, shoulder pads, VCR tape, vinyl record)
- Information cards about each object assembled by the teacher from reference materials or websites

Procedure

1. Begin to read *Imaginative Inventions*, stopping to discuss each. Then pose the question, "What are some other inventions you have used?" Some possible answers might be a toothbrush, hair dryer, or juice box. The key point is that these objects don't seem strange because students know the purpose of each.

2. Explain to the students that each team of two or three will be provided with a household object that isn't often used today but that their parents or grandparents may have used when they were young. The task is to come up with as many possible suggestions for what the object was used for in the time provided.

3. After five minutes, ask the students to switch objects with another team, thus allowing them to have another chance at developing their creative thinking. Have both groups of students combine and share their list of possibilities for the two different objects they had. Ask each group to explain the reasoning associated with their thoughts. Remind students that in an activity like this, there is no wrong answer and they should be supportive of one another. Last, student groups should come to a consensus on one answer and come up with a reason for the answer.

4. The student groups should then present their object to the class and provide the answer reached by consensus, along with their reasoning. When all groups have finished their presentations, the teacher should provide each group with the information card about their objects to check their answers and learn more.

5. Lead a discussion about when in the past these objects were used and why they were considered good inventions at that time. What purpose did they serve? Why were they better than something else? What objects do we

Imaginative Inventions

have that do the same thing today? Point out to students that all inventions have a life span during which they serve a specific purpose and then they are modified or something new is invented that will serve the same purpose in a better way. Ask them whether they can think of objects that have changed over time that still do the same thing (e.g., records to tapes to CDs to MP3 files).

6. Students can then brainstorm a list of objects they currently use that they think will become outdated or obsolete inventions years from now.

Grades 4–6: The Egg Drop

Purpose

Students will design and test a container that protects a raw egg from breaking when dropped.

Materials

- A collection of building supplies (egg cartons, string, plastic cups, paper towel, toilet paper rolls, Styrofoam packing peanuts, cotton balls, tape)
- Golf balls (one per team)
- Eggs (one per team)
- Resealable plastic bags

Procedure

1. Read the Internet Resources for detailed descriptions of the activity before beginning with the class. Read *So You Want to Be an Inventor* to the class, focusing on why the inventors developed or designed their particular device. For example, some inventions were designed to fill a need, such as the McCormick mechanical reaper.

2. Focus students' attention on the idea that some inventions are used to solve a particular problem and then pose the following problem for them to solve: "What type of container can you design to protect a raw egg from breaking when dropped from various heights?"

3. Ask the students what they have seen fall and not break. Sample answers might be pillows or stuffed animals. Have the students think about other types of containers that are meant to protect their contents from breaking (e.g., egg cartons or CD cases).

4. Explain the guidelines of the challenge to the students, which entails building a container to protect an egg from breaking when dropped from a set location. The teacher will need to identify a location from which to drop the containers (e.g., a stairwell or balcony). The teacher can start with a set height, such as 3 ft., then increase the height for future tries if they would like to make this a competition. Other guidelines for students are that they may only use the supplies provided by the teacher; they must provide a detailed drawing of their invention prior to building it; and they must conduct a trial with a golf ball first, then the raw egg. Teams should be made up of two to four students and are dependent on the classroom structure.

5. The design process will probably take several days to complete, with initial discussions led by the teacher (about force and motion). Students should then brainstorm what type of attributes the containers should have to be sturdy enough to withstand a fall from a set height while also protecting or cushioning an egg inside. After the brainstorming session, students can begin designing and labeling the parts of their containers. They can also start to support their reasons for their designs. Students can drop the golf balls into different materials to determine whether they absorb the impact. The students should consider possible substitutions of ideas

and trade-offs for their designs. After developing a final idea, they should begin to build their containers.

6. Once building is complete, students should do a trial test using the golf balls to see whether their containers hold up to the impact and force of hitting the ground. Students have one more opportunity to adjust their designs, allowing for slight modifications. Last, students test their designs using the raw eggs.

7. Finally student groups discuss what aspects of the design worked well and what aspects would be changed if they were to create a new model and why.

Reference

National Research Council (NRC). 1996. *National science education standards.* Washington, DC: National Academies Press.

Internet Resources

Egg Drop
http://scienceolmpiad.kulbago.com/Events%20 2010/Egg%20Drop%202010.pdf

Naked Egg Drop
http://sciencespot.net/Media/clubnkegg.pdf

The Egg Drop
http://books.nap.edu/openbook.php?record_ id=4962&page+162

Chapter 16

Let's Try It Out in the Air

By Emily Morgan and Karen Ansberry

Because we can't see it or hold it in our hands, many students tend to think of air as "nothing." The lessons here will challenge this fundamental misconception. Using two engaging trade books with "stop-and-try-it" formats, students participate in activities, collect evidence, and develop explanations to prove that air is indeed "something"—it is a gas that has weight and takes up space.

Trade Books

Let's Try It Out in the Air: Hands-On Early-Learning Science Activities
By Seymour Simon and Nicole Fauteux,
illustrated by Doug Cushman
Aladdin, 2003
ISBN 978-0-689-86011-9
Grades K–2

SYNOPSIS
Part of the Let's Try It Out! series, this book presents simple activities and explanations that teach young children about the properties of air.

I Face the Wind
By Vicki Cobb, illustrated by Julia Gorton
HarperCollins, 2003
ISBN 978-0-688-17840-6
Grades 3–6

SYNOPSIS
This book from Vicki Cobb's Science Play series uses a stop-and-try-it format to explain that air has weight and takes up space.

Let's Try It Out in the Air

Curricular Connections

The *National Science Education Standards* suggest that students experience science in a form that gives them the opportunity to construct their own ideas and explanations (NRC 1996). It is tempting to offer explanations and introduce vocabulary before students have experienced a scientific phenomenon. We often do this thinking that it will save time and allow us to move along to the next concept. But if students experience the phenomenon first, then they have a context to help them understand the concept they are learning—and a desire to know the scientific vocabulary to describe it. These two books pose questions that can be answered by doing activities that temporarily take the child away from the book. The teacher reads on after the child has made the discovery. This way, the book reinforces what the child has found through experience.

The *Atlas of Science Literacy* identifies the understanding that "air is a substance that surrounds us and takes up space" as a fundamental elementary school benchmark (AAAS 2001). In the K–2 lesson, students answer the question "Is air something or nothing?" In the 3–6 lesson, students learn through activities from the read-aloud that wind is moving air and that air has weight, takes up space, and therefore is matter. They learn about ways that wind can cause damage and also ways that we can benefit from wind.

Grades K–2: Air—Something or Nothing?

Materials

- Paper bag
- Sticky notes (two colors)
- Construction paper
- Materials in the book *Let's Try It Out in the Air*

Engage

Before class, draw a large question mark on the front of a paper bag and fold over the top so there is nothing but air inside. Show students the mystery bag and have them ask "yes or no" questions about what is inside. Then, open it up to show them the inside of the bag. Some students may say that the bag contains nothing. Some may say it contains air. Write "Is air something or nothing?" on the board, then make a three-column chart or bar graph with these choices: "Air is nothing"; I'm not sure"; and "Air is something." Give each student a sticky note (all the same color) and have them write their answers: "something," "nothing," or "not sure." Have them place their sticky notes in the corresponding columns on the chart or bars on the graph. Discuss the reasons students voted the way they did.

Explore/Explain

Tell students they will be reading and exploring to find the answer to the question they voted on. Show them the cover of *Let's Try It Out in the Air*. Explain that you will read parts of the book and then stop and try some of the activities. You can do some as demonstrations and others as student activities, depending on time and materials. After the read-aloud and activities, revisit the question, "Is air something or nothing?" Students should realize that air is "something." Have students share their evidence from the activities they tried. Evidence may include that air can (a) fill a balloon, (b) blow off a hat, (c) help airplanes fly, (d) fill up a paper bag, and (e) slow the fall of an object.

Elaborate

Tell students that good explanations are based on evidence. Throughout this lesson they collected a lot of evidence that proves that air is something. Tell them that as a class, they are going to create a bulletin board to prove that air is something. Create a bulletin board titled "Air Is Something: Here's Our Evidence." Make a list on the board of evidence that air is "something." Have students choose one piece of evidence, write it on a piece of construction paper, and illustrate it. For example, a student could write, "Air slows the fall of a feather," then draw a picture of a person dropping a feather. Display the bulletin

Let's Try It Out in the Air

board where others can see and where students might have the opportunity to explain their findings to others.

Evaluate

Give students a different color sticky note from what they had in the Engage phase of the lesson. Have them vote again on the question, "Is air something or nothing?" by writing "something" or "nothing" or "not sure" on the sticky note and placing it on the corresponding column in the chart or bar on the graph. Compare the two colors of sticky notes. Most likely, all students will now realize that air is "something." This chart is a good visual representation of how the students' ideas have changed based on the evidence they have collected. Explain that as we learn more about science and have more experiences, our ideas change. Tell them that it's okay if they did not have the right answer in the beginning. What is important is that they now know the correct answer and that this learning should be celebrated.

Grades 3–6: I Face the Wind

Materials

- Images of wind damage
- Materials from the book *I Face the Wind*
- Paper
- Pencils

Engage

In advance, do an internet image search on "wind damage," and choose some to share with your students. Be sure the photos are not disturbing or frightening to students. Ask them to look at the photos closely and infer what happened. Students may infer that there had been a storm, tornado, or hurricane. Ask them to explain the evidence that led them to their inferences. Tell students that all of the damage caused in the photos was done by wind. Ask students to talk to a partner about their own experiences with wind.

Explore/Explain

Ask students, "What is wind? Is wind matter?" Invite students to share their ideas on these questions. Then, show students the cover of *I Face the Wind*. Explain that you will read parts of the book and then try some of the activities. You can do some as demonstrations as well. Be sure to do the activity that proves that air takes up space (catching air in a large plastic bag) and the one that proves that air has weight (balancing the balloons on a hanger). After the read-aloud and activities, revisit the questions about wind. Students should realize through the activities and explanations in the book that wind is moving air and that air is matter.

Elaborate

Tell students that although extreme winds can cause damage, wind can be helpful. Show students the *Energy 101: Wind Turbines* video on the U.S. Department of Energy website (DOE; see Internet Resources). This video explains how turbines turn mechanical energy into electrical energy and shows several different types of wind turbines. Next, show students the DOE's animated Map of U.S. Wind Capacity that illustrates where wind turbines are located in the United States and how they have expanded since 1999. Ask students to infer why certain areas of the country have wind turbines and others do not. Then show them the Map of U.S. Wind Resources, which shows the average wind speeds in different parts of the country. Students should notice a correlation between amount of wind turbines (wind capacity) in an area and the average wind speeds (wind resources) in that area. It is important to mention that there are other factors that determine where turbines are placed, such as terrain, aesthetics, and population. Discuss whether students have seen wind turbines in your area. Click on your state on the Wind Resources map to find average wind speeds in your area and discuss the possible reasons why your state has turbines. (All of these maps are printable.)

Evaluate

Tell students that a friend of yours does not believe that air is matter. He thinks air is nothing because he cannot see it or hold it in his hands. He does not believe that air has any effect on him at all. Have students write a letter to this person providing evidence that air is matter. They must include one example that proves air has weight, one example that proves air takes up space, two examples of how air/wind can affect people in positive ways, and two examples of how air/wind can affect people in negative ways.

References

American Association for the Advancement of Science (AAAS). 2001. *Atlas of science literacy.* Washington, DC: AAAS.

National Research Council (NRC). 1996. *National science education standards.* Washington, DC: National Academies Press.

Internet Resources

Energy 101: Wind Turbines Video
www1.eere.energy.gov/windandhydro/wind_how.html

Map of U.S. Wind Capacity
www.eia.doe.gov/kids/energy.cfm?page=wind_home-basics

Map of U.S. Wind Resources
www.windpoweringamerica.gov/wind_maps.asp#us

Chapter 17

A Balancing Act

By Christine Anne Royce

Many children have seen circus performers walk a tightrope, and they gaze with awe at their skill and ability to keep their balance and not fall from great heights. The allure of walking a tightrope and the science of balance behind it is the focus of these activities. Through one fictional and one nonfictional tale, students will have the opportunity to explore activities that involve balance.

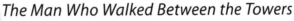

Trade Books

Mirette on the High Wire
By Emily Arnold McCully
Scholastic, 1993
ISBN 978-0-590-47693-5
Grades K–3

SYNOPSIS
A young girl named Mirette learns to walk along a high wire under the tutorage of the "Great Bellini." What she doesn't know is that he has lost his courage to perform, and through their trials together, both learn to conquer their fear.

The Man Who Walked Between the Towers
By Mordicai Gerstein
Scholastic, 2003
ISBN 978-0-439-70041-2
Grades 4–6

SYNOPSIS
The planning and execution of Philippe Petit's goal to walk a tightrope between the two towers in New York City is described through text and wonderful illustrations. Throughout the story the reader is told how Philippe and his assistants managed to get the gear to the roof, string the tightrope between the two towers, and perform before being arrested.

A Balancing Act

Curricular Connections

Balance is something that all children learn to do from the day they take their first steps. They continue to expand their physical abilities by climbing, running, and engaging in other childhood games. However, they don't necessarily think about balance when they are engaged in these activities.

The grades K–3 activity asks young students to put themselves in the role of young Mirette on the high wire as they try to walk a tightrope and change different variables to see how those variables affect their ability to stay above a certain point. Students are asked some basic scientific questions and use "their observations to construct reasonable explanations for the questions posed" (NRC 1996, p. 121). Furthermore, they also begin to see that "the position and motion of objects can be changed by pushing or pulling" (p. 127) (in this case with gravity) and begin to develop an intuitive sense of where their body should be placed.

Although the term *center of gravity* isn't developmentally appropriate for this age level, "teachers can build on the intuitive notions of students without requiring them to memorize technical definitions" (p. 126).

In the grades 4–6 activity, students will consider what factors go into balancing an object by manipulating variables such as position of the wire and amount of mass added. This opportunity to design a balanced object allows them to "develop general abilities, such as systematic observation, making accurate measurements, and identifying and controlling variables" (p. 145). Activities such as these are engaging and fun for students to investigate the concepts of balance and center of gravity.

Grades K–3: Learning to Balance

Materials

- Masking tape
- Four two-by-four blocks of wood (4–6 ft.)

Purpose

Students will engage in different physical activities in which they investigate ways to balance themselves.

Procedure

After reading *Mirette on the High Wire* to the class, ask the students what Mirette learned about balancing. Referring back to the pictures in the text and the cover will help students realize that she used her outstretched arms to assist with her balance.

Invite the students to become like Mirette and the Great Bellini and walk a "tightrope" in the classroom. Begin by placing a piece of masking tape across the floor. This pretend tightrope will allow students to safely practice putting their feet heel to toe on the "rope" with no risk of falling. Students should be reminded that if they step off the tape and onto the floor, they have "fallen" off the tightrope. Make sure the area is clear of hazards students could stumble into, such as table corners.

Ask students whether they have any suggestions that might make walking the tightrope easier, and allow them to try their ideas. Other students can make observations about what happens during the different attempts. Students will quickly see that the ability to keep their balance while walking heel to toe is more challenging than it looks.

Students should be invited to predict whether they think it will be hard or easy to stay directly on the masking tape. They should make other predictions as well: What do they think would happen if they walked the tightrope with books on their heads? Students can try to walk with their arms held out to their sides; with a book held with both hands overhead; with a book held in one hand with the other arm stretched out; or with a book held close to the body. All of these different positions will require students to adjust how they walk on the tightrope because their centers of gravity will change based on the position of the books and their arms. Although addressing the concept of *center*

A Balancing Act

of gravity at this age may not be developmentally appropriate, students can make predictions about what they think will happen before each attempt and then investigate the outcome. Asking students to describe why they think a certain position made it easier or harder will allow them to begin to develop the idea that balance can be affected by position of an object with different weight.

After students practice on the masking tape, make the task a bit more daring by lining the masking tape with two-by-fours, thus giving a little bit of height to the tightrope. Make sure the wood blocks are smooth without rough edges or other means for students to get slivers or scrapes. This activity should be done in the gym with floor mats lining both sides of the wood block path to prevent an injury should a student lose his or her balance and fall. Ask the students to initially repeat walking heel to toe and see whether they are able to keep their balance as Mirette did on a much higher wire.

Once students have had an opportunity to learn how to keep their balance, ask them about other phenomena or events that require balance. Some possible answers include students sitting on top of monkey bars, a bird sitting on a wire, or a squirrel running across the telephone line.

Grades 4–6:
Building a Balanced Object

Materials
- Pencils or craft sticks
- One piece of 20 in. pliable wire (20–24 gauge)
- Hexagonal nuts, washers, weight sets
- Safety glasses or goggles

Purpose
Students will attempt to make different objects balance while investigating the center of gravity.

Procedure
Read *The Man Who Walked Between the Towers,* then ask the students to look closely at the different positions that Philippe is in as he walks the high wire between the towers. The positions in the photos show that he keeps the tightrope directly below the center of his body whether standing up or lying down. Students may also notice that the pole he uses is also spread evenly across the wire.

Ask students why it matters that half of Philippe's body is on one side of the wire and half is on the other. Potential answers may relate to his weight being equally divided or perhaps that it helps him balance. Although it does help him balance, center of gravity must also be considered.

Explain to the students that they will be provided with a pencil or craft stick that they are being asked to balance on the tip of their index finger. Once students have the materials, ask them to balance the object and then describe what they found. Many of the students will state that it tips over too easily or that they can't balance it standing straight up. Some students will turn it sideways and get it to balance that way, which is fine and allows for a rich discussion. Refer them to the book and ask them what purpose they think the balancing pole plays in helping Philippe keep his balance. Allow them to pose possible answers.

Now that the students have tried to balance their pencil by itself, provide each student with a 20 in. piece of soft, pliable wire (20–24 gauge wire works best and can be obtained from home improvement stores). Students and teachers are to wear eye protection (safety glasses or goggles) when doing this activity. Remind students to be careful when working with sharp objects like wires, as they can puncture or scrape skin.

Have the students wrap the wire once or twice around the pencil so that it will stay in place. The best way to do this is by having the wire divided equally in half. However, based on the length of wire on each side and the different amount of weight on each side, students will be able to accomplish this task in different ways.

A Balancing Act

In addition to the wire, provide students with paper clips, different-size hexagonal nuts for bolts, rubber washers, metal washers, and if available, actual weights for use in a science lab. Students can add weight to each side of the wire to see what helps in balancing the pencil. Students should add different weights to each side of the wire and bend the wire to find a design that allows them to easily balance the vertically upright pencil on the tip of their finger.

Propose the following questions to the students and ask them to investigate: Does it matter where the wire is placed on the pencil—at the top, in the middle, or at the bottom? What happens when you add weight to one or both sides of the wire? What do you observe happening as you bend the wire downward or upward? How can you change the distribution of the weight to better balance the pencil? Where is the point of support for the object?

Have students consider what a point of support is. A *point of support* is the actual object that is supporting something or someone trying to balance on another object. For these designs the point of support is a student's finger, whereas in the book, the point of support is the rope itself.

Ask students to make observations about when their designs are balanced, which should be when the object is directly above or below the point of support. This is called the *center of gravity* (Figure 17.1).

Regardless of whether the students ever learn to be Mirette or Philippe and walk on a tightrope, activities such as these will help them understand the concepts of balance and center of gravity in everyday life.

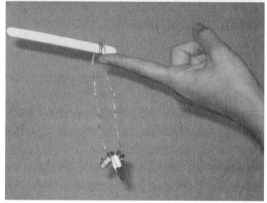

Figure 17.1. Finding Center of Gravity

PHOTOGRAPH COURTESY OF THE AUTHOR

Reference

National Research Council (NRC). 1996. *National science education standards.* Washington, DC: National Academies Press.

Chapter 18

Gravity

By Emily Morgan and Karen Ansberry

W hat goes up must come down … as long as you aren't floating in space, that is! So let's explore the concept of gravity. In the lesson for grades K–2, through several activities and readings, students learn that the force of gravity is what gives things weight. In the lesson for grades 3–6, students predict how different games and toys would work without gravity and then check their predictions by watching videos of the toys and games in action on the International Space Station (ISS) and the space shuttle.

Trade Books

I Fall Down

Written by Vicki Cobb, illustrated by Julia Gorton
HarperCollins, 2004
ISBN 978-0-688-17842-0
Grades K–3

SYNOPSIS

From the Vicki Cobb Science Play series, this book uses a stop-and-try-it format to explore the basics of gravity.

Floating in Space

By Franklyn M. Branley, illustrated by True Kelley
HarperCollins, 1998
ISBN 978-0-06-025433-9
Grades 4–6

SYNOPSIS

This book from the Let's-Read-and-Find-Out Science series describes how astronauts deal with weightlessness.

Curricular Connections

The *National Science Education Standards* suggest that early elementary students' understanding of force and motion concepts be developed primarily from manipulating objects and describing their motion (NRC 1996). *I Fall Down* poses questions that can be answered by doing activities such as dropping objects and observing them fall. After students explore, the book provides an explanation. Some key concepts include gravity is always pulling things down; as long as you're on Earth, you can't get away from gravity; and gravity gives things weight. In the upper elementary grades, the *Standards* suggest that students describe the specific forces affecting the motion of an object. To get students thinking about this, the lesson for grades 3–6 suggests that they try some simple toys and games. Students then predict how the toys and games would work in a weightless environment: the space shuttle or ISS. To check their predictions, the students watch online video segments of the toys being demonstrated in space by astronauts and cosmonauts in the space shuttle and ISS. One common misconception about the weightless environment in the space shuttle or ISS is that there is no gravity. In fact, it is Earth's gravity pulling on these spacecraft and everything inside that keeps them in orbit. As they orbit Earth, they are in a state of free fall, which makes everything feel weightless. This concept of the space shuttle and space station being in free fall during orbit might be too difficult for some elementary students to understand, so instead of explaining this, we simply refer to the people and objects in the space shuttle and ISS as experiencing "weightlessness."

Grades K–2: Gravity Pulls

Materials

- Assortment of balls
- Key
- Penny
- Block
- Jar of molasses or honey
- Spoon
- Dry sponge
- Small bar of soap
- Two identical rubber bands
- Child's shoe
- Adult shoe
- Bathroom scale

> Note: When choosing objects to drop, use only solids with minimal air resistance rather than things like feathers or tissues.

- Playground equipment, such as a slide
- Ball
- Drawing paper
- Crayons or markers

Engage

Tell students that on the count of three, you want them to jump as high as they can. Count, "1, 2, 3, jump!" Then ask, "Why do you think you always come down after you jump?" Have students turn and talk with partners and then share their ideas with the class. Tell students that they will be learning about why they always come back down after they jump up by reading a book and doing some fun activities.

Explore/Explain

Prior to reading *I Fall Down*, decide which activities you will stop and try during the read-aloud and have those supplies handy. Depending on the age of your students, you may decide to do some of the activities as demonstrations and others as whole-class activities. Show students the cover of *I Fall Down*. Ask them to signal when they hear the answer to the question "What makes things fall?" Students should signal when you read page 10, which says, "Know

Gravity

what makes things fall? It's a force called gravity." Continue reading the book aloud, stopping to try the different activities, then reading the explanations on the following pages.

Elaborate

Reread pages 10 and 11 of *I Fall Down*, which explain that as long as you are on Earth, gravity is always pulling things down. Tell students that you are going to take them to the playground to see how gravity affects them when they play. When you get to the playground, have students sit together where they can look at the playground equipment. Have a student go down the slide and ask, "What pulled him down the slide?" Have another student throw a ball high up in the air and ask, "What pulled the ball back down to the ground?" Continue to demonstrate gravity with other things on the playground, being sure to use the word *pull* so that students understand that gravity is a pulling force.

Evaluate

Back in the classroom, ask students to finish the sentence, "Gravity pulls …" and create an illustration to go with it. Examples include "Gravity pulls me down the slide"; "Gravity pulls syrup onto my waffle"; and "Gravity pulls my paper airplane to the ground." Display all of the sentences and pictures on a bulletin board titled "Gravity Pulls."

Grades 3–6: Toys in Space

Materials

- Video clip of hockey game
- Toys in Space videos (see Internet Resources)
- Toys in Space student page (p. 93)
- Various toys from the Toys in Space project, such as
  * Gyroscope
  * Magnetic Whee-Lo
  * Yo-yo
  * Jacks
  * Paper airplane
  * Paddleball
  * Slinky
  * Marbles
  * Spinning top
- Poster paper
- Markers

Engage

As a preassessment of your students' background knowledge of gravity, show them a clip of a hockey game and ask them how Earth's gravity affects the game of hockey. They should realize that gravity pulls the puck to the ice, holds the players down, and basically pulls everyone and everything toward the ground. Then ask, "How do you think the game of hockey would work in space?" Have them turn and talk to a partner. Tell students that astronauts wondered the same thing, so in 2002, they brought a hockey puck and hockey sticks to try out on the ISS. Show them the "International Toys in Space" video online that shows astronauts and cosmonauts in the ISS playing hockey (see Internet Resources). Discuss how the game worked differently in the weightless environment of the ISS from how it does on Earth, and how the astronauts and cosmonauts adapted the game to make it work better in weightlessness (e.g., they strapped their feet to the wall and put the puck over the vent).

Explore

Set up stations around the classroom with various toys from the Toys in Space project such as a gyroscope, a magnetic Whee-Lo, a yo-yo, jacks, a paper airplane, paddleball, a Slinky, marbles, and a spinning top. Give each student a copy of the Toys in Space student page (p. 93). They should play with the toy at each station, think about how Earth's gravity affects the way it works, and predict how it worked when astronauts tried it on the ISS or space shuttle. Have students share their predictions.

Explain

Show the video clips of astronauts using the same toys on the ISS or space shuttle. Students should record how the toys really worked in weightlessness

and compare them to their predictions. Next, read *Floating in Space* aloud, which explains more about weightlessness and how astronauts deal with it as they work, eat, and sleep. Have students listen for strategies and tools astronauts use to help them work in a weightless environment (e.g., Velcro, straps, bungee cords, and attaching things to the walls).

Elaborate

Have students choose one of the toys that didn't work well in space and write a paragraph explaining why it works differently in weightlessness. Then, have them write or draw their ideas on how the toy could be modified to perform better in weightlessness. They can incorporate ideas that the astronauts used in the book *Floating in Space*.

Evaluate

Have students choose one of their favorite toys or games and create a version of it that would work in weightlessness. They can create new rules or strategies, draw pictures of how it would work, and even give it a new name. Have them share their new toys or games with the rest of the class at a "Space Games" convention.

Reference

National Research Council (NRC). 1996. *National science education standards.* Washington, DC: National Academies Press.

Internet Resources

Resource Component for Toys in Space
http://aesp.nasa.okstate.edu/ftp/anderson/toysweb/index.htm

Toys in Space DVD
http://corecatalog.nasa.gov/item.cfm?num=009.0-11D

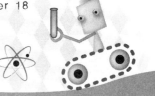

Name: _____

Toys in Space

Name of Toy _____

Play with the toy for a few minutes. How does Earth's gravity affect this toy?

_____

_____

_____

How do you think the toy would work differently on the International Space Station or space shuttle?

_____

_____

_____

Name of Toy _____

Play with the toy for a few minutes. How does Earth's gravity affect this toy?

_____

_____

_____

How do you think the toy would work differently on the International Space Station or space shuttle?

_____

_____

_____

Name: _____

Toys in Space

Name of Toy _____

Play with the toy for a few minutes. How does Earth's gravity affect this toy?

_____

_____

_____

How do you think the toy would work differently on the International Space Station or space shuttle?

_____

_____

_____

_____

Name of Toy _____

_____

Play with the toy for a few minutes. How does Earth's gravity affect this toy?

_____

_____

_____

How do you think the toy would work differently on the International Space Station or space shuttle?

_____

_____

_____

_____

Chapter 19

Roller Coasters!

By Karen Ansberry and Emily Morgan

Students of all ages are fascinated by the ups, downs, loops, and twists of roller coaster rides! What they may not realize is that there is a lot of science involved in making a roller coaster work. Put students in the shoes of a roller coaster designer as they work in teams to create their own roller coasters.

Trade Books

Roller Coaster

By Marla Frazee
Sandpiper, 2006
ISBN 978-0-15-205744-2
Grades preK–4

SYNOPSIS

From waiting in the long line to zooming down a hill and zipping through a loop, Frazee's illustrations capture the excitement of a little girl's first experience on a roller coaster.

Roller Coaster! Motion and Acceleration

By Paul Mason
Raintree, 2006
ISBN 978-1-4109-2616-8
Grades 4–6

SYNOPSIS

This exciting book puts the reader in the shoes of a roller coaster designer. Full-color photographs, bolded words, and insets reveal the forces that affect roller coaster motion.

Curricular Connections

The *National Science Education Standards* recommend that students explore and describe motion by pushing, pulling, throwing, dropping, and rolling everyday objects (NRC 1996). The Standards suggest that K–4 students begin to focus on the position and motion of objects as well as the motion and forces required to control the objects. By making careful observations and recording data, even students in the earliest grades can begin to look for patterns in their work with motion and can determine the speed of an object as fast, faster, or fastest. In the K–2 lesson, students investigate how to control the speed of a model roller coaster and discover how forces affect the motion of objects. The Standards suggest that in grades 5–8, students continue to have concrete experiences with simple objects but begin to describe the forces acting on these objects. In the 3–6 lesson, students learn how the forces of gravity and friction affect the motion of roller coasters.

Grades K–2: Fast, Faster, Fastest

Materials (per group of three to four students)

- 6 ft. length of foam insulation to fit a 1 in. pipe (split lengthwise; available at home improvement stores) with a plastic 20 oz. cup taped to the end
- A ball that will roll in the split pipe insulation prepared above (e.g., foosball, large marble, ball bearing)
- Paper
- Crayons or markers

Engage

Show students the cover of *Roller Coaster* by Marla Frazee and ask, "Have you ever been on a roller coaster? What was it like? If you haven't, what do you think it would be like?" Begin reading the book, but stop after reading pages 14 and 15, where the roller coaster is slowly going up the hill. Ask, "What

do you think the next picture in the book will look like?" Continue reading the book, but stop after reading page 27 ("Wheeeeeee!"). Have students close their eyes and imagine what it would feel like to be on the roller coaster in the book. Ask, "How would you feel if you were on this roller coaster?" For fun, have the whole class make their "roller coaster faces" on the count of three. Then, finish reading the book aloud.

Explore

Announce to students that they are going to work with partners to design their own roller coasters.

Show the foam track-and-cup setup and the ball to students and tell them these represent a roller coaster. Caution them not to throw the ball or push it down the track. They should simply release the ball and let it roll. Show them the cup and ask, "What do you think the cup is for?" (to catch the ball or to stop the ball from rolling away).

Give each pair of students the supplies and have them complete the following challenges:

- Can you make the ball roll from one end of the track and stop in the cup?
- Can you make the ball roll faster? Slower?
- Can you make the ball go over one hill on your roller coaster? Two hills?
- Can you make the ball go through a loop on your roller coaster?

Encourage students to make observations about where on the track the ball moves fastest and slowest as they explore.

Explain

Bring students back together and ask these questions:

- How did you make the ball roll faster? (by raising one end a lot higher than the other)
- How did you make the ball roll slower? (by raising one end only a little higher than the other)

Roller Coasters!

- Which was the highest, the first hill or the second hill? (the first hill had to be the highest to get the ball going fast enough to go over the second hill)
- How did you make the ball go over the hills or around loops on your roller coaster? (by making the beginning of the track steep)

Elaborate/Evaluate

Ask students to draw roller coasters using what they have learned from their models. They can make as many hills and loops on their track as they wish, as long as they think the roller coaster would actually work in real life. Have them label where the roller coaster would be moving the fastest and slowest. Ask, "What do you need at the beginning of the ride to get the car moving fast?" (a high hill). "Can a second hill be higher than a first hill?" (no, the first hill has to be the highest). Evaluate students' understandings about motion by asking, "Where on the track does your roller coaster car move the fastest?" (toward the bottom of the hills). "Where on the track does your roller coaster car move the slowest?" (toward the top of the hills). "If you were a roller coaster designer in real life, what tips would you give about roller coaster design?" (The first hill should be the highest. Make sure there is a high hill before a loop. Don't make the loop too big, etc.).

Grades 3–6: Designers' Challenge

Materials (per group of three to five students)
- Four 6 ft. sections of foam pipe insulation
- Roll of duct tape
- Foosball
- Index cards

Engage

Show students the cover of the book *Roller Coaster!* by Paul Mason. Read aloud pages 4–5 (including insets) and ask, "What do you think it would be like to be a roller coaster designer? What qualifications

Figure 19.1. Making a Roller Coaster

PHOTOGRAPH COURTESY OF THE AUTHORS

do you think a roller coaster designer would need?" (know a lot about science and math, have a college degree, be able to think creatively, etc.). Tell students that a roller coaster designer's job is to design the most exciting ride possible while making sure the ride is safe. Two important things the designer must consider are the materials and space available.

Explore

Show students a foosball and a 6 ft. section of pipe insulation with a cup taped to the end as described in the previous lesson. Tell students that these are the materials they will use to make a model roller coaster. Their task is to use these materials to find out as much as they can about roller coaster design. Give students specific challenges as described in the K–2 lesson during exploration time.

Explain

Have students explain their observations about roller coaster motion (e.g., how did they make the ball go faster, slower, and so on). Then tell students that the book *Roller Coaster!* explains the science behind roller coaster design. As you read pages 6–27 (including insets), have students listen for the meanings of the following key words (write these on the board): *force, direction, gravity, speed,* and *friction.* Stop after reading about each of these words and ask, "What does this mean for you as a roller coaster designer?" (There needs to be enough speed to get the roller coaster over a hill; if the tubing is

too narrow, friction could slow the roller coaster down too much; and so on.)

Elaborate

Tell students that they will be working with a team of three to five students to design roller coasters. They will use the information they learned in the book and their discoveries from the Explore phase to help with their designs. Review the importance of the key words to a roller coaster designer. Each team will be given four 6 ft. sections of tubing, a roll of duct tape, and a foosball. They may use chairs, desks, walls, and anything else in the classroom to help support their roller coaster. The teacher and the other students will rate each roller coaster in two areas: fun and safety. Tell students that roller coaster designers spend the majority of their time drawing designs and making models on their computers. So, before they can start building, teams must plan the roller

coaster on paper. After they turn in a plan and the teacher approves it, they can start building. Have students test their designs and revise them as necessary. Next, they will present their roller coasters to the rest of the class. Have students write down a score for fun and a score for safety for each roller coaster on index cards. Students must also include an explanation of their scores and ideas for improvement. After the presentations, allow teams time to read the cards and try out some of the suggestions.

Evaluate

Have each member of the team write an explanation of how forces affect the motion of the designed roller coaster. Their explanations must include the correct use of the key words.

Reference

National Research Council (NRC). 1996. *National science education standards.* Washington, DC: National Academies Press.

Chapter 20

Secrets of Flight

By Karen Ansberry and Emily Morgan

The date was December 17, 1903. The place was a windswept beach near Kitty Hawk, North Carolina. With Orville Wright at the controls and his brother Wilbur running alongside, the plane took off. This event lasted only 12 seconds, but it made history as the first successful sustained flight by a powered, heavier-than-air, piloted aircraft. The Wright brothers had uncovered the secrets of flight. Help students explore the history of flight and use problem-solving skills to improve the flight distances and flight times of paper gliders.

Trade Books

Animals in Flight
By Steve Jenkins and Robin Page
Sandpiper, 2005
ISBN 978-0-618-54882-8
Grades K–4

SYNOPSIS
Jenkins's trademark cut-paper collages illustrate the spare, large-print text, which reveals how, when, and why animals have taken to the air. Smaller-print text gives more details and explanations. A two-page spread gives a brief introduction to the history of human flying machines.

How People Learned to Fly
By Fran Hodgkins, illustrated by True Kelley
HarperCollins, 2007
ISBN-13 978-0-06-445221-2
Grades 4–8

SYNOPSIS
This Let's-Read-and-Find-Out Science book reveals the many obstacles that have been overcome in the history of human flight. The four forces of flight (gravity, lift, drag, and thrust) are explained with simple text and illustrations.

Curricular Connections

The lessons described here address some physical science concepts, such as motion and forces, but the true focus of the lessons lies in the areas of Science and Technology and History and Nature of Science. The *National Science Education Standards* (*NSES*) indicate that students in grades K–8 should develop the abilities of technological design; namely, to identify a problem or need, design a solution, implement a solution, evaluate a product or design, and communicate the design process (NRC 1996). They advise that students should be involved in activities that are meant to meet a human need, solve a problem, or develop a product. In both of these lessons, students use the technological design process to build simple gliders, test them, and make changes to improve them. The NSES also suggest that students view science as a human endeavor by learning about the men and women who have contributed to science and technology throughout history. In the K–3 lesson, students discuss the need for human flying machines and how people throughout history have worked hard to solve the problems of flight. Students in grades 4–6 explore the history of innovation in human flying machines and share their ideas for what the future holds for human flight.

Grades K–3: Animals in Flight

Materials

- Paper airplane template (See Internet Resource)
- Markers or crayons for decorating the gliders

Engage

Show students the cover of *Animals in Flight* and ask them to spend one minute, working with a partner, on brainstorming a list of animals that fly. As you list the animals on the board, ask, "What do all these animals have in common?" (all have wings or winglike flaps). As you read *Animals in Flight* aloud, have students signal when any animals from their list are mentioned. After reading, have students recall additional flying animals from the book. Ask students to think about the advantages of flight for animals (e.g., escaping danger, finding and catching food, and moving from place to place). Next, ask why people might have tried so hard throughout history to invent flying machines (e.g., getting places faster, the thrill of flying like a bird, and the challenge of doing what some thought impossible).

Explore/Explain

Ask students, "Do you ever look at birds and imagine what it would be like to fly? Do you ever run with your arms out pretending that you are flying?" Tell them that if they answered yes to either of these questions, they are not alone. People have dreamed of flying for thousands of years. Then ask, "Do you know who built the first airplane that flew successfully?" (Orville and Wilbur Wright). Tell students that these two brothers spent a lot of time observing birds in flight and designed gliders based on their observations. If you can, take your students outdoors to observe birds in flight. Students should notice that birds sometimes flap their wings and sometimes glide with their wings outstretched. Have students flap their outstretched arms to feel the air pushing back against their hands. Explain that pushing against air helps make flight possible. Next, give each student a free downloadable template for a beginner paper airplane (see Internet Resource). After reviewing safety procedures (i.e., do not throw paper airplanes toward people), have students test their gliders. Then ask, "How is your glider like a bird?" (It has wings and can glide on the air.) "How is it different?" (It can't fly by itself, you have to throw it, the wings don't flap, and so on.)

Elaborate

Tell students that Orville and Wilbur Wright were never completely satisfied with their airplane designs. They were always trying to improve their safety, flight time, distance traveled, and so on. Brainstorm ways to solve the following problem:

Secrets of Flight

How can you make a paper glider fly farther? For example, students can throw the gliders with a different amount of force or at a different angle, place a paper clip on the nose of the glider, cut small flaps on the wings or tail, or even try an entirely new paper airplane template. Mark a start line on the ground with tape or chalk. Have students throw the gliders with no modifications first and mark with tape or chalk where they land. Then, have them make one change to their glider designs and throw them again. Be sure to explain that to see if a certain change helps a glider fly farther, they should only change one thing at a time. Have students repeat this problem-solving process until they are satisfied with their glider designs. Explain that the Wright brothers were successful because they didn't give up—they kept trying different designs until they solved the problems they were working on.

Evaluate

Have each student decorate and name his or her glider that traveled the longest distance and describe the features of the glider that made it travel the farthest. (There is a paper clip on the nose, this one had the longest wings, it has a flap on the tail, and so on.)

Grades 4–6: Forces of Flight

Materials

- Flight Data Sheets (pp. 103 and 104)
- Stopwatch
- Paper airplane template (see Internet Resource)
- Forces of Flight student page (p. 105)
- Paper, pencils, and markers for timelines

Engage

Throw a paper glider and ask students, "What are some terms you think of when you hear the word flight?" Give groups of two to three students sticky notes and ask them to write one word about flight

on each, and then sort the sticky notes into categories. Write the word *flight* in the center of a sheet of chart paper and circle it. Discuss the students' categories and terms, and then have students place their notes in the form of a semantic map. Display the map throughout the lesson, adding new categories and terms as students learn more about flight (Figure 20.1).

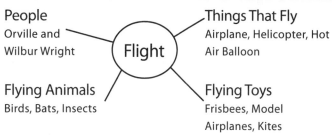

Figure 20.1. Sample Semantic Map for Flight

People
Orville and
Wilbur Wright

Flying Animals
Birds, Bats, Insects

Flight

Things That Fly
Airplane, Helicopter, Hot
Air Balloon

Flying Toys
Frisbees, Model
Airplanes, Kites

Explore

Tell students that the Wright brothers, who invented the first successful airplanes, began by building gliders. Give each pair of students Flight Data Sheets, a stopwatch, and a paper airplane template (see Internet Resource). Find an open space in your classroom or hallway. After reviewing safety procedures (i.e., do not throw paper airplanes toward people), have students take turns throwing the airplane while their partners use stopwatches to determine how long the airplanes are in the air. Next, have students go outdoors, record the weather conditions, repeat the timing procedure, and answer the remaining questions on their Flight Data Sheets (pp. 103 and 104).

Explain

Have students discuss the answers to the questions on the Flight Data Sheets. Students should understand that the flight times differed because of the following factors: the person throwing it, the force of the throw, the person timing it, weather conditions, and so on. Next, give each student a copy of the Forces of Flight student page (p. 105). Tell students that as you read *How People Learned to Fly*, you would like them to listen for each of the

Secrets of Flight

four forces (lift, thrust, drag, and gravity). Read the book aloud, stopping to discuss how each force affects flight. Then, have students place the words in the correct places on the student page. Ask, "What creates lift on your paper airplane?" (the wings). "What creates drag on your paper airplane?" (air resistance). "What creates thrust with your paper airplane?" (the force we throw it with). "What force pulls your paper airplane to the ground?" (gravity).

Elaborate

Ask students how they might use what they have learned from the book and their labeled diagrams to solve the following problem: How can you improve the flight time of your paper glider? Brainstorm ideas such as placing a paper clip on the nose of the glider, cutting small flaps on the wings or tail, or even trying an entirely new paper airplane template. Provide time for students to modify and test their glider designs. Next, have students create posters that include labeled diagrams of their most successful designs, the steps of their design processes, and the results of their trials. They should also describe why they think those designs worked best, using the four forces of flight (lift, drag, gravity, and thrust) in their explanations.

Evaluate

Since the Wright brothers' first flight in 1903, there have been many innovations that have improved human flight. Have students create a timeline that details the evolution of human flight since 1903 and include a prediction for the future: What will the next big innovation be, and when will it occur? Have students display their timelines in the classroom and have a class discussion on what students predict for the future of human flight.

Reference

National Research Council (NRC). 1996. *National science education standards*. Washington, DC: National Academies Press.

Internet Resource

Fun Paper Airplanes
www.funpaperairplanes.com

Name: _____

Flight Data Sheet

Indoor Flight

1 Find an open space in your classroom or hallway. Safety warning: DO NOT THROW PAPER AIRPLANE TOWARD OTHERS. One person should throw the paper airplane while the other measures the amount of time the paper airplane stays in the air. Record the time in the table below.

Flight Data: Indoors

Trial	Flight Time (sec.)	Flight Observations
1		
2		
3		
4		

2 Were all of your plane's flight times the same? Why or why not?

_____

_____

_____

Outdoor Flight

3 With your teacher, go outdoors. Describe the weather conditions. _____

_____

_____

4 Repeat the procedure above with your airplane outdoors, and record your data on the following page.

Name: _____

Flight Data Sheet

Flight Data: Outdoors

TRIAL	FLIGHT TIME (SEC.)	FLIGHT OBSERVATIONS
1		
2		
3		
4		

5 Were all of your plane's flight times the same? Why or why not?

_____

_____

_____

6 Did your paper airplane glide longer inside or outside? Why do you think that happened? _____

_____

_____

7 How do you think weather conditions affect a real airplane's flight?

_____

_____

_____

8 Why doesn't your paper airplane keep flying like a real airplane?

_____

_____

_____

Forces of Flight

WORD BANK
Lift
Thrust
Drag
Gravity

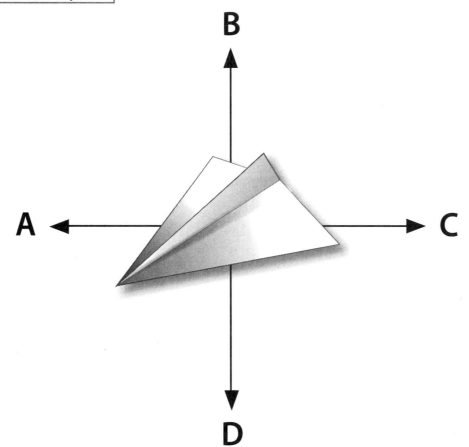

Chapter 21

Flick a Switch

By Karen Ansberry and Emily Morgan

When students flick on lights, boot up computers, or turn on televisions, do they think about how that energy is produced? The majority of electricity in the United States is generated from power plants that burn fossil fuels, causing large amounts of carbon dioxide and other pollutants to be released into the atmosphere. Conserving energy not only saves us money, but it also helps reduce the demand for fossil fuels. The activities here focus on how electricity gets to our homes, the reasons electricity should be conserved, and practical ways that we can use energy wisely.

Trade Books

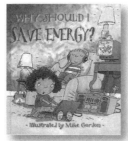

Why Should I Save Energy?
By Jen Greene, illustrated by Mike Gordon
Barron's Educational Series Inc., 2005
ISBN 978-0-7641-3156-1
Grades K–4

SYNOPSIS
Through simple text and fun illustrations, a young boy explains to his friend why her family should save energy, what it would be like at home and school without energy, and how we all can conserve energy.

Flick a Switch: How Electricity Gets to Your Home
By Barbara Seuling, illustrated by Nancy Tobin
Holiday House, 2003
ISBN 978-0-8234-1729-2
Grades 4–6

SYNOPSIS
Informative text and cartoonish illustrations describe how electricity was discovered, how early devices were invented to make use of electricity, how electricity is generated in power plants, and how electricity is distributed for many different uses.

Flick a Switch

Curricular Connections

The *National Science Education Standards* (*NSES*) suggest that in grades K–4, students should learn that electricity in circuits can produce light, heat, sound, and magnetic effects (NRC 1996). In the K–2 lesson, students spend time identifying items in their homes that are powered by electricity, and then determine what the electricity in each circuit does. They also talk with their parents or guardians about where the electricity in their homes comes from. In addition, the *NSES* suggest that K–4 students understand that the supply of many resources is limited and that resources can be extended through decreased use. Students learn that electrical energy is often produced by burning fossil fuels—a limited resource—and they devise ways to conserve energy at school and at home. The lesson for grades 3–6 delves deeper into how electricity is produced and carried into our homes. Students also learn about ways to conserve electricity and create a plan for conserving it in their schools and homes. This lesson addresses several parts of the Science in Personal and Social Perspectives strand of the *NSES*, such as understanding the benefits and risks of technology, identifying causes of environmental degradation, and realizing the ability to contribute to solving a problem.

Grades K–2: Saving Energy

Materials

- Electricity at Home take-home page (p. 111)
- Paper and markers for posters

Engage

Turn the light switch off in your classroom and then turn it back on. Ask students, "What makes the lights turn on when you flick the switch?" (electricity). Next, ask, "Where do you think the electricity comes from?" (answers will vary). Then ask, "What can electricity do besides make light?" (Students may have observed that electricity can produce light, heat, or sound. They may not be

aware that it can also produce a magnetic force, which in turn can cause a motor to run. For K–2 students, it is sufficient to say that electricity can also make things move.) Discuss student responses, and then tell students that they will find out where electricity comes from during the next few days.

Explore

Have students work with parents or guardians to complete the Electricity at Home take-home page (p. 111). They will make a list of as many things they can find in one room of their homes that use electricity, identify what the electricity does in each thing (makes it light up, heat up, make sounds, or move), and then find out where the electricity to power their homes comes from.

Explain

Have students discuss the items in their homes that use electricity and explain to one another what they found out about where the electricity for their homes comes from. Then explain to students that electricity is a form of energy. Energy makes change; it does work for us. It heats and lights our homes, plays songs on the radio, moves cars along the road, and bakes cookies in ovens. Show them the cover of *Why Should I Save Energy?* Before reading, ask why they think we should save energy. Then read the book aloud, stopping on page 17 when the boy asks, "What do you think would happen if our homes ran out of energy?" Invite students to turn to a partner and respond to this question. Read on, stopping at pages 19 and 21, which ask what would happen if the school and town ran out of energy, providing time for students to discuss their answers with their partners.

Elaborate

After reading, show students the cover again and ask them to explain the answer to the question in the title, "Why should I save energy?" Students should be able to explain that energy should be saved because we don't want to run out of energy

Flick a Switch

and because saving energy would save our families money. Also, to make electricity, power plants often have to burn fuel such as coal, which creates air pollution. So when we save energy, we create less pollution. Ask students to recall some ideas from the book for saving energy, such as turning the lights off when you don't need them, and closing windows and doors when the heat is on. As a class, make a list of ways to save energy at school. Based on the list, create some signs to post around the school like "Don't forget to turn off the lights" or "Ride the bus to school instead of a car."

Evaluate

Have students create and illustrate posters to hang at home that describe to their families at least two reasons to save energy (e.g., save their family money and reduce pollution) and two ideas for ways to save energy (e.g., turn the TV off when no one is watching and turn the lights off when you leave a room).

Grades 3–6: Using Electricity

Materials

- Electricity at Home take-home page (p. 112)
- Let's Learn About Electricity anticipation guide (p. 114)

Engage

Ask students to think back to when they woke up that morning. Ask, "In what ways did you use electricity between then and now?" Have them turn to a partner and discuss all the things they used that were powered by electricity between the time they woke up to that moment. Say, "It's very easy to use electricity, but have you ever thought about how much you use and where it comes from?" Then discuss the following questions:

- Where does the electricity in your home come from?
- How does it get there?

- What does it cost?
- How is your household electricity use measured?

Explore

Give students a copy of the Electricity at Home take-home page (p. 112), which asks them to make observations of their electric meter, find out the name of their electric company, investigate the different ways their families use electricity, and imagine what a day would be like without electricity. Have students share their answers on the take-home pages the following day.

Explain

Tell students they are going to find out more about electricity. Give them the Let's Learn About Electricity anticipation guide (p. 114) and have them fill in the "before reading" column. Next, introduce the author and the illustrator of *Flick a Switch: How Electricity Gets to Your Home*. Ask students to signal when they hear the answers to any of the true/false statements from the anticipation guide. After reading, ask students to fill in the "after reading" column of the anticipation guide and discuss the correct answers (1. False; 2. True; 3. True; 4. False; 5. True; 6. True). Then turn back to pages 14–15, which explain how power plants produce electricity from water, steam, wind, nuclear power, and fossil fuels. Explain that although electricity itself is a clean source of energy, power plants create various forms of pollution as a result of generating electricity. Then challenge students to find out how the electricity they use in their own homes is generated (e.g., steam, wind, nuclear power, or fossil fuels). Most power companies have websites that provide detailed descriptions about how they generate electricity. Have students explain how their families' power companies generate electricity.

Elaborate/Evaluate

Tell students there is a movement toward energy conservation in the United States and that everyone can help conserve energy. Explain that conserving

energy simply means reducing unnecessary energy use and waste. Conserving electricity can reduce pollution and save money. Challenge your class to start an energy-saving campaign in your school. They can create posters, write announcements, and videotape commercials. The U.S. Department of Energy has a website for young people called Lose Your Excuse (see Internet Resources). Students can visit the site to print energy-saving posters, play games, watch public service announcements featuring kids, and download an energy conservation action plan. They can also visit the Kids Saving Energy website for more ideas and activities (see Internet Resources).

Reference

National Research Council (NRC). 1996. *National science education standards.* Washington, DC: National Academies Press.

Internet Resources

Kids Saving Energy
www.eere.energy.gov/kids

Lose Your Excuse
www.loseyourexcuse.gov

Flick a Switch

Name: _____

Electricity at Home

Electricity is a form of energy. We depend on electricity for many things at home. With an adult helper, make a list of all the items in one room of your home that use electricity. Place a check mark in each box that describes what electricity makes it do.

Room _____

ITEM THAT USES ELECTRICITY	MAKES LIGHT	MAKES SOUND	HEATS UP	MOVES

Talk with your adult helper about where the electricity for your home comes from.

What did you find out? Draw or write on the back of this page.

Electricity at Home

1. With an adult helper, locate your electric meter at home. An **electric meter** measures the amount of electricity your family uses in your home. Electricity is measured in units of power called **watts**. The amount of electricity we use over a period of time is measured in **kilowatt-hours**, or the energy of 1000 watts for one hour. Many electric meters have a spinning disk below the dials that speeds up as you use more electricity. Your electric company reads this meter to find out how many kilowatt-hours your family uses in a month, then calculates your electric bills.

Location of electric meter: _____

Drawing of electric meter:

Flick a Switch

2. Name of company that provides electricity to your home:

    _____

3. List the ways your family uses electricity at home in each category.

CATEGORY	ITEMS THAT USE ELECTRICITY
Heating and cooling	
Lighting	
Food storage and preparation	
Entertainment	
Other	

4. Imagine what a day in your life would be like without electricity. Write about how it would be different.

_____

_____

_____

_____

_____

_____

_____

Let's Learn About Electricity

Anticipation Guide

Before **True or False**		*After* **True or False**
_____	1. Ben Franklin invented electricity.	_____
_____	2. Lightning is a form of electricity.	_____
_____	3. Electricity can be made with a magnet and coiled wire.	_____
_____	4. Electric power is carried from a power plant to your town by trucks.	_____
_____	5. Wires carry electric current through the walls of your house.	_____
_____	6. One-third of the people in the world live without electricity.	_____

Chapter 22

The Wonder of Water

By Karen Ansberry and Emily Morgan

Water is an extraordinary substance that we often take for granted. Not only is it what makes our planet uniquely habitable, water is the only substance on Earth that naturally occurs in three different forms. In these activities, students will explore some of water's fascinating properties.

Trade Books

I Get Wet

By Vicki Cobb, illustrated by Julia Gorton
HarperCollins, 2002
ISBN 978-0-688-17838-3
Grades preK–4

SYNOPSIS

This book from the Science Play series poses several questions that can be answered by doing simple activities with household items. It features a young boy who discovers some of the properties of water by pouring it into different containers, observing it drip from a faucet, and sliding it on waxed paper. Bold illustrations, lively text, and creative use of typography help highlight the captivating properties of water.

A Drop of Water: A Book of Science and Wonder

By Walter Wick
Scholastic Press, 1997
ISBN 978-0-590-22197-9
Grades 4–6

SYNOPSIS

Spectacular color photographs show water in its various forms—droplets, ice cubes, steam, snowflakes—while concepts such as condensation, capillary action, and surface tension are explained through simple text.

The Wonder of Water

Curricular Connections

Young children begin their study of matter by observing and describing objects and their behavior. The important but abstract ideas about matter all begin with simple observations (NRC 1996). In the K–3 lesson, students make observations of the adhesive and cohesive properties of water. They learn that water sticks to itself and some other things. They also learn that water does not stick to things that are greasy or waxy, then apply that knowledge to explain why ducks don't get wet when they are in the water.

It's tempting in the upper elementary grades to introduce atoms and molecules to explain phenomena such as adhesion and cohesion, but according to the *National Science Education Standards*, the introduction of atoms and molecules is premature for these students and can distract from the understanding that can be gained from observation and description of how matter behaves. So for grades 3–6, the focus of the lesson is still on developing ideas about matter based on observation, but the lesson becomes more sophisticated with the introduction of the scientific terms *adhesion* (water being attracted to other things, such as paper towels, glass, etc.) and *cohesion* (water being attracted to itself). Students make observations of water's properties by completing the activities at seven water exploration stations, learn about adhesion and cohesion from nonfiction reading and teacher explanation, and then use these terms to develop and revise their explanations of the phenomena they observe.

Grades K–3: Sticky Water

Materials

- Clear plastic cup
- Water
- Pennies
- Feathers (one per student)
- Petroleum jelly
- Water droplet cutout (p. 126)
- Materials for activities in *I Get Wet*

Engage

Fill a clear glass with as much water as it will hold before spilling. Then ask your students, "How many pennies can this glass hold before any water runs down the side?" (Answers will vary.) Try it! Students will be surprised to see just how many pennies the glass can hold. Close up, they will be able to observe a dome of water forming at the top of the glass. Say, "How do you think this is possible? Let's read a book called *I Get Wet* to find out."

Explore/Explain

Vicki Cobb suggests that the best way to use her book is to do the activities described in the book as they come up during the reading. Before you begin reading, make sure you have all the necessary supplies at hand. Cobb also suggests not turning the page to the explanation until after the child has made the discovery. That way, the book will reinforce what the child has discovered through experience. After reading page 16, ask students why the glass of water could hold so many pennies. (The surface of the water acts like a skin.)

Elaborate

After reading the rest of the book (stopping to try the activities suggested), ask students what the book said about why ducks don't get wet. (A duck's feathers are coated with a kind of grease, and water doesn't wet grease.) Then, give each pair of students a sterile, dry feather (available at craft stores) and a small container of water. Have them predict what the feather will look like after it has been dipped in water. Try it! Next, give each pair another clean, dry feather and have them cover it lightly with petroleum jelly to simulate how a duck covers its feathers with grease. Have them predict what the feather will look like after it has been dipped in water. Try it! Students will discover that the clean feather gets very wet, but the greased feather does not absorb as much water.

The Wonder of Water

Evaluate

Create a "Water Can …" bulletin board by giving each student a water droplet cutout on which they complete the sentence, "Water can …" They may write any of the properties of water that they learned from the lesson, such as water can flow, make droplets, stick to itself, slide off waxed paper without wetting it, roll off a duck's back, stick together, move up a paper towel, and so on. Make a picture of a watering can to post on your bulletin board and staple each student's droplet under the spout.

Grades 4–6: Water Exploration Stations

Materials

(for demonstration)

- 2 clear plastic cups
- Water
- String
- Bin or bucket to catch water
- Food coloring (optional)

(per pair of students)

- 2 small plastic cups
- Water
- 50 cm string
- Bin or bucket to catch water
- Paper towels
- Materials for water exploration stations (See pp. 119–125)

Engage

Show students two clear plastic cups, one containing water and the other empty, and a wet 50 cm piece of string. Hold the cups about 25 cm apart. (Do not tell the students that the string is wet. Later, they will figure out that the wet string is the secret to the trick in the "Elaborate" section of this lesson.) Tell them you are going to attempt to pour the water from one cup, down the string, and into the other cup without moving the cups closer together. You may want to add food coloring to the water so students can see it flow down the string. Try it! Ask students how they think you did it. They may say that you had a special kind of string, a special kind of water, a hole in the middle of the string, and so on.

Explore

Tell students they are going to do some explorations with water that might help them figure out how you poured water down the string. Set up water exploration stations around the room (see pp. 119–125) that include some of the activities photographed in Walter Wick's *A Drop of Water*.

Explain

After completing the stations, have students discuss their observations and possible explanations for the phenomena they observed. Next, introduce the author and illustrator of *A Drop of Water* and begin reading. Ask students to look and listen for information that might help explain the things they observed in the water exploration stations. Stop after reading about each of the experiments performed at the stations and discuss how the new information can help them develop/revise their explanations. From the stations and the reading, students should conclude that water droplets are attracted to each other. Tell students that this property of water is called cohesion. Students should also conclude that water is attracted to other things like paper towels, plastic, glass, and so on. Explain that this property of water is called adhesion.

Elaborate

Tell students that they are now going to take what they have learned about cohesion and adhesion and try to pour water down the string. Give each pair of students one cup of water, one empty cup, a dry string, and a bucket or bin to catch any falling water. Let them experiment with the supplies. Through experimentation, students will soon discover the "secret" of getting it to work: The string must be wet! Move around the room and help pairs until all have been successful in pouring the water down the string.

Evaluate

Have students write thorough explanations of what causes water to flow down the string, using the words *adhesion* and *cohesion*: For example, "Water sticks to the string because of its adhesive property. Adhesion counteracts the force of gravity pulling the water down. Water particles stick to each other because of their cohesive property, causing the water to flow without breaking apart. So, the water is sticking to the string (adhesion) as well as the water that is already on the string (cohesion)."

Reference

National Research Council (NRC). 1996. *National science education standards.* Washington, DC: National Academies Press.

Water Exploration Stations
Station 1: Water on a Penny

Materials:

- Penny
- Pipette
- Glass of water

1. How many drops of water do you think will fit on the top of a penny? _____
2. Try it!
3. Record your results.

TRIAL	NUMBER OF DROPS
1	
2	
3	

Look at the water and penny at eye level from the side.
Draw and write your observations below.

Station 2: Molecules in Motion

Materials:

- Glass of water
- Blue food coloring
- Pipette

1. What do you think will happen if you put two drops of food coloring in a glass of water without stirring it? _____

   _____

   _____

   _____

2. Try it! (Be sure the water is very still before adding the food coloring and do not bump the table as you watch.)

3. Draw and write your observations below.

Station 3: Water on Waxed Paper

Materials:

- Waxed paper
- Glass of water
- Pipette

1. Place two drops of water close together on waxed paper. What do you think will happen when you push one drop toward the other with the pipette? _____

   _____

   _____

   _____

2. Try it!

3. Draw and write your observations below.

Station 4: Soap and Water

Materials:

- Penny
- Pipette
- Glass of water
- Clean paper clip
- Soapy paper clip

1. Put 20 drops of water on a penny. What do you think will happen if you put the tip of the soapy paper clip gently in water that's on the penny? _____

_____

_____

2. Try it!
3. Draw and write your observations below.

[ ]

4. Put 20 drops of water on a penny. What do you think will happen if you gently put the tip of the clean paper clip in the water that's on the penny?

_____

_____

_____

5. Try it!
6. Draw and write your observations below.

[ ]

Station 5: Wet Paintbrush

Materials:

- Dry paintbrushes
- Glass of water

1. Observe a dry paintbrush. What do you think the bristles will look like when you hold the paintbrush in the water?

2. Try it!

3. Draw and write your observations below.

4. How do you think the bristles will look when you pull the paintbrush out of the water?

5. Try it!

6. Draw and write your observations below.

Station 6: Bubbles

Materials:

- Glass of water
- Bubble solution
- Two bubble wands

1. Is it possible to blow a bubble with plain water? _____
2. Try it!
3. Draw and write your observations below.
4. Is it possible to blow a bubble with soapy water?_____
5. Try it!
6. Draw and write your observations below.

WATER	SOAPY WATER

Station 7: When Water Flows Up

Materials:

- Pipette
- Waxed paper
- Glass of water colored with blue food coloring
- Drinking straw
- Coffee straw
- Strips of paper towel

1. Is it possible for water to flow up? _____

2. Try it! Place 10 drops of water on a piece of waxed paper. Hold the end of a drinking straw in the water.

3. Draw and write your observations below.

4. Now, hold a coffee straw in the water.

5. Draw and write your observations below.

6. Now, hold the end of a strip of paper towel in the water.

7. Draw and write your observations below.

DRINKING STRAW	COFFEE STRAW	PAPER TOWEL

The Wonder of Water

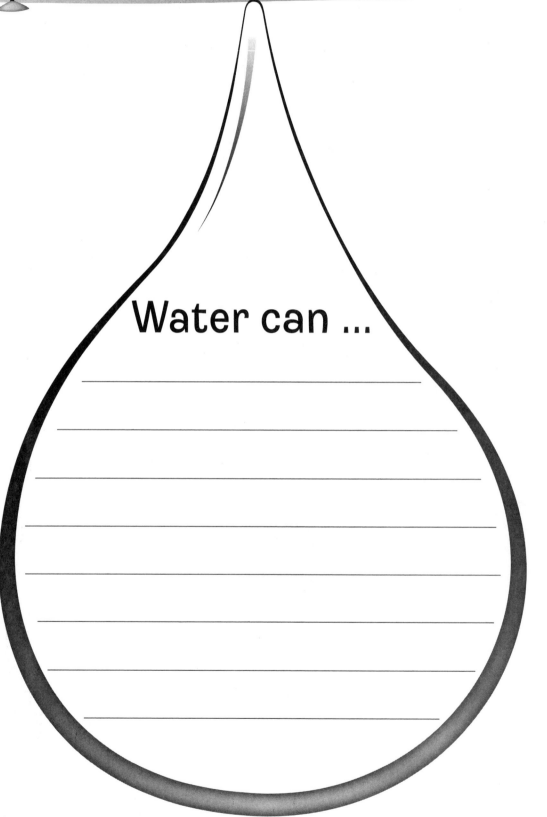

Water can ...

_____

_____

_____

_____

_____

_____

_____

Chapter 23

Kitchen Chemistry

By Christine Anne Royce

The kitchen is a wondrous place for children to make observations and explore the basics of chemistry. Seize the opportunity and help students build process skills while cooking or baking. Almost everything we eat and certainly everything that is combined in the kitchen has some basis in the physical sciences—from mixtures to solutions to the glorious tastes of food.

Trade Books

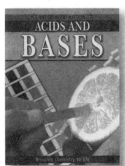

Pancakes for Breakfast
By Tomie dePaola
Sandpiper, 1978
ISBN 978-0-15-670768-8
Grades K–3

SYNOPSIS
This book has no words to read, but it offers wonderful illustrations and pictures of a woman who wants to make pancakes but lacks several of the key ingredients. The story unfolds through the carefully created images of what the woman does to locate the ingredients and make the pancakes.

Acids and Bases: Why Chemistry Matters
By Lynette Brent
Crabtree Publishing, 2008
ISBN 978-0-7787-4246-3
Grades 3–6

SYNOPSIS
An overview of acids and bases is presented for young readers, providing a handy reference.

Curricular Connections

For many reasons, chemistry is not a concept often considered for elementary classrooms—whether it be the perceived developmental appropriateness of the topic or our own memories of what chemistry may have been like when we learned it during our academic careers. However, simple chemistry concepts are part of the physical sciences, which students need to understand the natural world. According to the *National Science Education Standards,* "children's natural curiosity leads them to explore the world by observing and manipulating common objects and materials in their environment. … physical science in grades K–4 includes topics that give students a chance to increase their understanding of the characteristics of objects and materials that they encounter daily" (NRC 1996, p. 123). This idea is incorporated into the first activity as students make observations and illustrate or sequence the events that occur during the process of making pancakes. They then get to consider what they observed and determine whether the change was chemical or physical.

As students get older, they need additional opportunities to experiment and ask questions while building science-process skills. According to the *NSES,* "In grades 5–8, students observe and measure characteristic properties … of pure substances and use those properties to distinguish and separate one substance from another" (NRC 1996, p. 149). The activity for older students allows them to measure a specific characteristic of substances as they determine whether the substance is an acid or a base and its approximate location on a pH scale.

Grades K–3: Choosing a Change

Materials

- Pancake ingredients*
  * one egg
  * 1.25 cups buttermilk
  * 2 tbsp. vegetable oil
  * 1.25 cups flour,
  * 1 tbsp. sugar,
  * 1 tsp. baking powder,
  * 1 tsp. baking soda,
  * 0.5 tsp. salt
  * butter
  * electric griddle or access to a stove

*You may need to increase the recipe depending on the class size.

Purpose

Students will make observations and determine whether the changes they observe represent a physical or a chemical change.

Procedure

Begin by introducing *Pancakes for Breakfast* and asking students, "What does making pancakes have to do with science?" Students will most likely say they use the stove to cook pancakes or that ingredients are mixed together to make pancakes. Share the book with the class, stopping at each page and asking for comments about what is happening in the pictures. For example, students may observe, "The woman takes out a bowl and flour." Based on this picture, they may also infer, "… to make pancakes." Ask students to predict what happens next before turning the page.

Introduce the two different vocabulary terms—*physical change* and *chemical change*. Students at this age will not have the requisite background necessary to determine the difference. Matter can undergo either a physical or a chemical change. Physical changes happen when some aspect such as form, shape, or appearance of the matter changes, but the original matter is still the same substance. An example of a physical change would be water that freezes, melts, or evaporates. The water is still water but in a different form. Chemical changes happen when substances that have been combined change into a new substance that contains new properties. After discussing the terminology, present to the

Kitchen Chemistry

class the following examples and have them classify them as a physical or a chemical change:

1. Newspaper that has burned and results in ash: chemical change

2. Baking soda that is mixed with vinegar: chemical change

3. Piece of wood that is chopped into smaller pieces for firewood: physical change

4. A large sheet of paper torn into smaller parts: physical change

If possible, mix up a batch of homemade pancakes in front of the class using the recipe from the book (ingredients on previous page). Keep students at a safe distance while making the pancakes as a teacher demonstration. While making the pancakes, have the students make a storyboard or sequencing chart that allows them to describe each step that goes into making pancakes. By separating the process into parts, students are better able to identify individual events and determine which type of change occurs. Examples include measuring the flour, cracking the eggs open, and mixing the buttermilk into the flour. While demonstrating the process involved in making pancakes, ask students questions such as "When I measure the flour, am I changing the substance? Why or why not?" This connects the students' observation skills with conceptual thinking about what they know about physical and chemical changes. The teacher can also ask questions that require the students to recall what happens next in the story.

Students are able to make observations of events, infer if they are physical or chemical changes, and support their reasoning with evidence. Have them recount the process, noting which steps are physical changes and which are chemical changes. (All of the actions in the book are physical changes with the exception of the cooking of the pancake itself.)

As a follow-up, repeat this activity with another children's favorite, *Stone Soup* (Brown 1997). In this story, although the vegetables put into the soup get softer, it is still a physical change that occurs.

Grades 4–6: Determining pH

Materials
- Safety goggles
- Test tubes or small plastic cups
- Eyedroppers
- Head of red cabbage
- Pot
- Water
- Grater
- Distilled water
- Glass jar
- Baking soda
- White vinegar
- Ammonia
- Milk
- Lemon juice
- Milk of magnesia
- Carbonated beverage
- Chart paper

Purpose
Students will determine whether common household substances are acids or bases on the pH scale.

Procedure
Use the book *Acids and Bases* as a reference and read the section on what an acid is and what a base is. One of the concepts addressed in the book is acids and bases and testing for their pH. The activity presented here allows students to use common household materials to begin to investigate acids and bases. Materials are classified on a scale ranging from 1 (strong acid) through 14 (strong base) with a 7 being considered neutral.

Prepare a cabbage indicator ahead of time. To do this, grate a head of red cabbage, place the grated cabbage into a pot, cover with water, and bring to a boil. Let it boil for 20–30 minutes or until the water turns dark purple. Once the liquid has been cooled, strain out the grated cabbage pieces, and discard. The indicator liquid can be stored in a sealed glass jar in the refrigerator for up to three days.

Have students determine what happens when an indicator solution is used with known substances. Place 25 ml of distilled water in each of three test tubes or small plastic cups and number. In cup 1, place a few grams of bicarbonate and mix; in cup 2, leave only the distilled water; and in cup 3, place 5 ml of white vinegar. Have the students place a few drops of the cabbage indicator into each cup and note the color change. Cup 1 is a base (pH higher than 7), cup 2 is neutral (pH of 7), and cup 3 is an acid (pH lower than 7). The juice will turn a pinkish color in acidic solutions and a greenish color in basic solutions. Throughout this process, ask students to make observations of what they see.

After students are comfortable with testing known solutions, provide them with some un-knowns to determine whether they are acidic or basic. Again, label a set for each group from one through five. Using the chart (Figure 23.1)

for teacher reference, place the correct liquid in each tube.

After they don safety goggles, have students drop the same number of drops (approximately five) into each container and observe the color change. Students then determine whether each is an acid or base using the cabbage indicator. Discuss the group results as a class. Then pose the following question: "Can you determine which is the most acidic to the most basic based on the color of the indicator liquid?" Students will be able to do so as the color of the indicator deepens based on the pH level.

Provide the approximate pH level for each substance tested and discuss how accurate the cabbage indicator was. The higher the pH number, the more basic the substance, thus the more purple to green to yellow the color. In contrast, the lower the pH number, the more acidic the substance is and the indicator will reflect a more pinkish color.

Regardless of the chemistry concept explored, using common kitchen ingredients makes classroom activities easier for the teacher to set up and more interesting for the student to engage with.

References

Brown, M. 1997. *Stone soup*. New York: Aladdin.

National Research Council (NRC). 1996. *National science education standards*. Washington, DC: National Academies Press.

Figure 23.1. Chart for Teacher Reference

TUBE	SUBSTANCE	DETERMINATION	APPROX pH
#1	15 ml ammonia	Base	11.0
#2	15 ml milk	Acid	6.6
#3	15 ml lemon juice	Acid	2.2
#4	15 ml milk of magnesia	Base	10.5
#5	15 ml carbonated beverage (soda)	Acid	2.5

Chapter 24

Secrets of Seeds

By Karen Ansberry and Emily Morgan

*F*rom a tiny radish seed to a giant coconut, seeds come in a multitude of shapes and sizes. They all share one amazing secret: the potential to grow into a new plant when conditions are right. These activities encourage students to observe a variety of seeds, match seeds to the plants they grow into, explore what seeds need to germinate and grow, and design investigations with seeds.

Trade Books

Seeds

By Vijaya Khisty Bodach
Capstone Press, 2007
ISBN 978-0-7368-9623-8
Grades K–4

SYNOPSIS

Simple text and bold, close-up photographs present the seeds of different plants, how they grow, and their uses.

Seeds

By Ken Robbins
Atheneum Books for Young Reader, 2005
ISBN 978-0-689-85041-7
Grades K–4

SYNOPSIS

Ken Robbins's stunning photographs and straightforward text explain how seeds grow and how they vary in size, shape, and dispersal patterns.

Curricular Connections

The *National Science Education Standards* state, "during the elementary grades, children build understanding of biological concepts through direct experience with living things, their life cycles, and habitats"(NRC 1996 p. 127, 1996). Specifically, the Standards state that early elementary students may not understand the continuity of life from seed to plant, but they can observe that the offspring of plants closely resemble the parent plants (NRC 1996). Students in grades K–4 should also learn that organisms have basic needs and can survive only if these needs are met. In the K–2 lesson, students observe seeds, match them to the fruits from which they came, and discover that seeds need water and warmth to germinate.

Identifying questions that can be answered through scientific investigations and designing and conducting investigations are fundamental abilities that should be learned in grades 5–8 (NRC 1996). In the 3–6 lesson, students not only match seeds with their corresponding fruits or vegetables using a chart key, but also design investigations to answer questions that they generate about seeds.

Grades K–2: Needs of Seeds

Materials

- Avocado pit
- Whole avocado
- Toothpicks
- Water
- Clear cup
- Various fruits (two of each, one for retrieving seeds)
- Container filled with potting soil
- Paper towel
- 1 gal. resealable plastic bag
- Measuring cup
- Seeds with a short germination time, such as cucumber or radish

Engage

Show students an avocado pit without identifying it. Pass it around the room and have them make observations and inferences about it. Then slice a whole avocado fruit in half to reveal the pit. Explain that it is a seed that grows into a tree that produces a fruit called an avocado. This particular fruit has only one seed. Some fruits have many seeds. Clean off the avocado pit and then stick four toothpicks into its sides to a depth of about 5 mm. Set the toothpicks on the rim of a container of water so that the pit is half-submerged in the water with its pointed side up. Change the water every other day. In a few weeks, students will be able to observe roots coming out of the bottom and leaves growing out of the top.

Explore

In advance, purchase a variety of fruits such as apples, pumpkins, melons, plums, and cherries, and retrieve some of the seeds from them. Leave one of each fruit whole. Give each pair of students an assortment of seeds from the fruits. Ask them if they think all of these seeds will grow into the same kind of plant (no). How do they know? (because the seeds are different sizes and shapes). Next, ask them to predict what kind of plant each seed will become. After they have had time to make predictions, give them the whole fruits and have them try to match up the seeds with the fruits.

Explain

Have students explain how they determined which seed would grow into which kind of plant. Then, open one of each fruit so students can see the seeds. Ask them to check their matches and correct any they had wrong. Next, ask students what they think needs to happen for the seeds to grow into plants. Show them the book *Seeds* by Vijaya Khisty Bodach. Ask students to listen for what seeds need to grow as you read aloud (soil, water, and warmth). They will also find out that birds, wind, and people help spread seeds so they can grow.

Secrets of Seeds

Elaborate

For this part of the lesson, you will need a container filled with potting soil, a paper towel, a 1 gal. resealable plastic bag, a measuring cup, water, and some seeds with a short germination time, such as cucumber or radish. Have students recall the needs of seeds from the book (soil, water, and warmth). Say, "I am wondering if seeds really do need soil to *germinate,* which means start to grow. What would happen if we planted some of these seeds on a wet paper towel?" Have students make predictions, then plant several seeds in the pot and the same number of seeds on a dry, flat paper towel inside a resealable plastic bag. Next, ask students if it would be fair to water the seeds in the pot but not the seeds on the paper towel (no). Explain that it is important to keep the experiment "fair" by giving all the seeds the same amount of water. Pour the same amount of water into both the container and the baggie. Place them near a sunny window or other warm place in the classroom. After several days, both will likely have germinated. Ask, "Do seeds need soil to *germinate?*" (no). Have students make daily observations. After a few weeks, the seeds in the soil will appear more healthy than the seeds on the wet paper towel. Students can then conclude that seeds, once they have germinated, grow best in soil.

Evaluate

Ask each student to bring in a seed and write a riddle about what the seed grows into. The riddle should include three clues, the needs of the seed, and the question "What am I?" For example,

- I am shaped like a teardrop.
- I grow in clusters on an ear.
- If you get me too hot, I will pop!
- I need soil, water, and warmth to grow. (Or, "I need water and warmth to germinate.")
- What am I?

Students can attach the seed to the top of the page, and then write the answer at the bottom and cover it with a liftable flap of paper.

Grades 3–6: Seed Investigations

Materials

- Variety of seeds (sweetpea, cherry, plum, watermelon, wheat, corn, acorn, maple, impatiens, and avocado)
- Hand lenses
- Seeds Chart Key (p. 135)
- Sentence strips
- Markers

Engage

Give each group of 2–4 students a variety of the following seeds featured in the book *Seeds* by Ken Robbins: sweetpea, cherry, plum, watermelon, wheat, corn, acorn, maple, impatiens, and avocado. Ask them to observe the seeds with a hand lens, measure them, and then predict what kind of plant each will grow into.

Explore/Explain

Give students a copy of the Seeds Chart Key (p. 135). Introduce the chart key as a special tool that scientists can use to identify unknown objects. Direct students' attention to the column headings and the pictures. Tell students that the first thing they should do when using a key is to look at any pictures, labels, or headings. This information will help them use the key to identify their objects. Model using the key to find the name of a seed. Students should then try to identify each seed using the key. When they find the name, they can place the seed in the appropriate row in the first column of the key. Then, read the book *Seeds* by Ken Robbins aloud to the class. Each seed is named and pictured in the book, so students can check their answers as you read.

Elaborate

Next, ask students what they are wondering about seeds. Ask each pair of students to write

a seed question on a sentence strip. Collect all of the sentence strips and read the questions aloud to the class. Explain that the type of investigation a scientist does depends on the questions he or she asks. As a group, sort the students' questions into "researchable questions" that can be answered using reliable sources of scientific information, and "testable questions" that can be answered by observing, measuring, or doing an experiment (see Figure 24.1).

Figure 24.1. Sample Questions Sort

Researchable Questions

- How do squirrels find the acorns they bury?

- What is the world's largest seed?

- What is the difference between a fruit and a vegetable?

Testable Questions

- How long does it take for an avocado seed to sprout?

- Do larger watermelons have more seeds than smaller watermelons?

- Will popcorn kernels from the grocery store grow into popcorn plants?

Evaluate

Have each group of two to four students select one of the testable questions and discuss ways to investigate the question. After designing their investigation and collecting data, they can brainstorm ways to communicate their results (pictures, data tables, graphs, etc.) and then share their findings with the class.

Reference

National Research Council (NRC). 1996. *National science education standards.* Washington, DC: National Academies Press.

Seeds Chart Key

PLACE SEED HERE	NAME OF SEED	COLOR	SHAPE	SIZE	SPECIAL CHARACTERISTICS
1.	Sweetpea	Brown	Spherical	< 0.5 cm	——
2.	Cherry	Brown or tan	Oblong	≥ 0.5 cm	Bumpy
3.	Plum	Brown or tan	Oblong	≥ 0.5 cm	Bumpy
4.	Watermelon	Brown or black	Oblong	≥ 0.5 cm	Flat
5.	Wheat	Tan	Oblong	≥ 0.5 cm	——
6.	Corn	Any	Irregular	≥ 0.5 cm	——
7.	Acorn	Brown, tan, or green	Irregular	≥ 0.5 cm	Cap present
8.	Maple	Tan or green	Irregular	≥ 0.5 cm	Wings present
9.	Impatiens	Brown or tan	Oblong	< 0.5 cm	——
10.	Avocado	Brown or tan	Spherical or oblong	≥ 0.5 cm	——

Spherical   **Oblong**   **Irregular** *(may be any shape)*

Chapter 25

Pumpkins!

By Karen Ansberry and Emily Morgan

Walk through any elementary school in the fall and you are bound to see students learning about pumpkins. Kids of all ages are fascinated by these versatile, edible members of the gourd family. Pumpkins are interesting and inexpensive and can be found in a variety of colors, shapes, and sizes, so they're terrific for using in hands-on math and science lessons. The activities described here use two picture books about pumpkins as well as some real pumpkins to engage students in the processes of scientific inquiry.

Trade Books

Pumpkin Circle: The Story of a Garden
By George Levenson, photographs by Shmuel Thaler
Tricycle Press, 2002
ISBN 978-1-58246-078-9
Grades K–2

SYNOPSIS
Poetic text and photographs follow a pumpkin patch as it grows and changes, from seeds to plants to pumpkins ready to harvest, to jack-o-lanterns and then to seeds again.

How Many Seeds in a Pumpkin?
By Margaret McNamara, illustrated by G. Brian Karas
Schwartz and Wade Books, 2007
ISBN 978-0-375-84014-2
Grades K–4

SYNOPSIS
Charlie and his classmates investigate to see which pumpkin has the most seeds: the smallest or the largest.

Curricular Connections

According to the *National Science Education Standards*, "students at all grade levels should have the opportunity to use scientific inquiry and develop the ability to think and act in ways associated with inquiry," (NRC 1996, p. 105). The following lessons use pumpkins as a springboard into inquiry. The lesson for grades K–2 focuses on making observations, asking questions, and designing a simple investigation. In the lesson for grades 3–6, students learn that some questions are best answered through investigation and others through research.

Grades K–2: Pumpkin O–W–L

Materials

- Small pumpkin in a brown, paper bag
- Wide variety of pumpkins
- Poster paper
- Markers

Engage

Before students enter the classroom, hide a pumpkin in a brown paper bag. Tell the class that you have something special in the bag and that you will be asking them to make some scientific observations about the "mystery object." Remind students that an observation is information that they get through their senses. An observation describes how something looks, feels, sounds, smells, and tastes (but in science class it is not safe to taste!). Invite a student to put his hand in the bag and make some observations of how it feels, without inferring or guessing what it is. Invite another student to hold the bag to feel how heavy it is. Ask another student to close her eyes and smell inside the bag. Cut out a small hole in the back of the bag and ask a student to describe the color. Write all of the students' observations on the board. Then have the class guess what the mystery object is.

Explore

Create a three-column O-W-L (Observations-Wonderings-Learnings) Chart about pumpkins on poster paper. Bring in several different varieties, sizes, colors, and shapes of pumpkins for students to explore. Divide students into groups and give each group a pad of large sticky notes. Tell students that throughout this lesson, they are invited to write their observations of their pumpkins on the sticky notes (so they can easily be placed onto the chart). You may want to cut open one of the pumpkins so that students can experience what the inside of a pumpkin feels like. After students have had sufficient time to explore the pumpkins, have them post their observations in the "O" column of the pumpkin O-W-L.

PHOTOGRAPH COURTESY OF THE AUTHORS

Pumpkins!

Explain

Show students the cover of *Pumpkin Circle*. Tell students that good readers asks themselves questions while they read. Tell them you are going to share with them some of your questions while you read just to let them know what you are wondering. As you read, share some wonderings, such as, "I wonder why the author called this book *Pumpkin Circle*. I wonder what it would feel like to touch the inside of a pumpkin. I wonder if all pumpkin seeds sprout two fat green leaves. I wonder if small seeds grow small pumpkins. I wonder if larger pumpkins have more lines on them than smaller pumpkins do." After reading, ask students to use their sticky notes to record some of their own wonderings about pumpkins from the reading and from their observations. Invite them to post their wonderings in the "W" column of the Pumpkins O-W-L chart.

Elaborate

Discuss ways that students could find answers to some of these questions. Students should understand that some questions can be answered by making observations or doing an experiment and others are best answered by reading a book, searching the internet, or asking an expert. Choose a simple, testable question from the wonderings column of the O-W-L Chart and guide students through an investigation to find the answer. For example, "Do larger pumpkins have more seeds than smaller pumpkins do?" could be answered by cutting open the pumpkins and counting and graphing the number of seeds inside.

Evaluate

Have students post their learnings about pumpkins in the "L" column of the Pumpkin O-W-L Chart. For example, "Pumpkins have slimy seeds inside," "Pumpkin plants have flowers," "Pumpkins come in many shapes and sizes," and "Pumpkins aren't always orange."

Grades 3–6: Pumpkin Investigations

Materials

- Several pumpkins of different sizes
- Newspaper
- Bowls
- Spoons
- How Many Seeds in a Pumpkin? student page (p. 141)

Engage

Hold up a pumpkin and pose the question, "How many seeds do you think are in this pumpkin?" Have students make some estimates. Then read aloud *How Many Seeds in a Pumpkin?* stopping at page 5 where the teacher asks the class, "How many seeds in a pumpkin? Does anybody know?" Then ask students how they could find out. Students should suggest opening up the pumpkins and counting the seeds.

Explore

Provide several pumpkins of different sizes for students to explore. In advance, cut the tops off of the pumpkins. Divide students into groups of three or four and give each group a pumpkin and a copy of the How Many Seeds in a Pumpkin? student page (p. 141). Ask them to collect the data in the table and then predict which pumpkin they think has the most seeds. Tell students that the goal for the day is to get all of the seeds out of the pumpkins. It is difficult to count the seeds when they are covered in slime. They will count the seeds the following day after they have dried out. Cover student work areas with newspaper and provide a large bowl and a spoon for scooping out the seeds. After the student areas are cleaned up and all of the seeds are in the bowls, read pages 6–13 of *How Many Seeds in a Pumpkin*, where the teacher says, "Tonight your homework is to think about how we should count all the seeds." Tell students that tonight you will be

drying the seeds and that *their* homework is to think of a quick, efficient way to count the seeds.

Explain

The next day, give students time to discuss their ideas on the most efficient and accurate way to count the pumpkin seeds. Then, read pages 13–19 where the students decide to count the seeds by twos, fives, and tens. Suggest that students may want to consider grouping their seeds by twos, fives, or tens. After students have had time to count their seeds, have them write the total number of seeds on the outside of their pumpkin with permanent marker. Display the pumpkins side by side so that students can see the results. Read the rest of *How Many Seeds in a Pumpkin?* Ask students how their results compare to those from the reading. In the book, the smallest pumpkin happens to have the most seeds, but your students' results may be different. Students learn from the book that for every line on a pumpkin, there is a row of seeds inside and that pollination, pumpkin variety, and time on the vine determine how many lines are on a pumpkin—and how many seeds are inside.

Elaborate

Next, ask students what other questions they have about pumpkins. Ask each pair of students to write down a pumpkin question. Collect all of the questions and read them aloud to the class. Explain that the type of investigation a scientist does depends on the questions he or she asks. As a group, sort the students' questions into "researchable questions" that can be answered using reliable sources of scientific information and "testable questions" that can be answered by observing, measuring, or doing an experiment, such as "Do larger pumpkins have larger seeds than smaller pumpkins do?"

Evaluate

Have each group of three to four students select one of the testable questions and discuss ways to investigate the question. After investigating the question, they can brainstorm ways to communicate their results (pictures, data tables, graphs, poster presentations, etc.), and then share their findings with the class.

Reference

National Research Council (NRC). 1996. *National science education standards.* Washington, DC: National Academies Press.

Pumpkins!

Name: _____

How Many Seeds in a Pumpkin?

1. Record the following data for your pumpkin:

PUMPKIN NAME (Name your own.)	DESCRIPTION	WEIGHT	CIRCUMFERENCE	NUMBER OF LINES

2. Observe the other pumpkins in your classroom. Which one do you think has the most seeds? Why? _____

    _____

    _____

3. Record the class pumpkin data below:

PUMPKIN NAME	WEIGHT	CIRCUMFERENCE	NUMBER OF LINES	NUMBER OF SEEDS

4   Circle the pumpkin that had the most seeds.

Chapter 26

Flower Power

By Christine Anne Royce

Summer means flowers in bloom! Each flowering plant produces a unique bloom that provides opportunities for students to make observations about plants. By comparing and contrasting flowers, students can connect their learning to the larger picture that all organisms have different structures that help them survive.

Trade Books

Planting a Rainbow
By Lois Ehlert
Harcourt Brace and Company, 1988
ISBN 978-0-15-305476-1
Grades K–2

SYNOPSIS

In this book, students are introduced to the planting cycle for various ornamental flowers. When bulbs should be planted, when seeds should be ordered, and when to go purchase seedlings are all explained as a young child and mother plant a rainbow of colorful blossoms.

The Reason for a Flower
By Ruth Heller
Puffin, 1999
ISBN 978-0-698-11559-0
Grades 2–5

SYNOPSIS

Ruth Heller explains the reason for a flower, shows pollinators visiting flowers in colorful illustrations, describes how seeds travel from place to place, and covers other interesting facts about flowers.

Curricular Connections

Connecting flowers to the curriculum can occur in two different areas. First, students at both the K–3 and 4–6 levels can continue to develop their "understanding of biological concepts through direct experience with living things, their life cycles, and their habitats" (NRC 1996, p.127). By making observations of and interacting with flowers, students can begin to understand the key ideas found in the selected children's literature—that plants grow in different ways and that flowers have different parts that help them reproduce. In *The Reason for a Flower*, the author helps introduce older students to the idea that "each plant or animal has different structures that serve different functions in growth, survival and reproduction" (NRC 1996, p. 129).

The second area that connects to the curriculum is that of process skills and conducting different kinds of investigations. The *NSES* emphasize that "scientists use different kinds of investigations depending on the questions they are trying to answer....includ[ing] describing objects, events, and organisms" (NRC 1996, p. 123). Older students do this as they dissect plants to learn about their structures.

Grades K–3: Observing Flowers

Purpose

Students will make observations of various flowers and describe them in qualitative and quantitative ways.

Materials

- An assortment of plants similar to those in *Planting a Rainbow* (Obtain the live plants and their source: seeds or bulbs.)
    * flowering bulbs (e.g., lily, iris)
    * seedlings (e.g., poppies)
    * seeds (e.g., marigolds, zinnias)
- Containers
- Potting soil
- Magnifying lenses

Some plants are poisonous or have saps that can cause skin irritation, so have students wear gloves. Make sure students thoroughly wash hands after handling plants. Remind young children to never put unknown plants in their mouth.

Procedure

1. In advance, plant some of your seeds and bulbs so you'll have some growing plants for students to observe. Save some of the unplanted bulbs and seeds for students to observe.

2. Read *Planting a Rainbow* to the class. While reading, have the students make observations of the illustrations. Ask questions such as, "What do you notice happens as a bulb germinates?" (something green starts to come out of the dirt) and "What comparisons can you make between seedlings, seeds, and bulbs?" (seedlings already have leaves, but seeds don't).

3. Show students the seeds, bulbs, and seedlings that you collected, and ask them to make observations about each. Next, ask them to compare and contrast the three objects. They may make observations such as, "Sometimes seeds are smaller than bulbs." The teacher may choose to cut open a bulb and seed and to uproot a seedling to allow the students to make observations about each in more detail.

4. Next, introduce the students to the plants that were started in advance. The plants should be grouped into three categories (grown from bulb, seed, and seedling). Allow students to make observations. These can be quantitative, with focus questions such as, "How many flowers are produced on each stalk?" or, "How many petals does each flower have?" They can also be qualitative, with focus questions such as, "What do you notice about the shape of the leaf?"

Flower Power

5. Allow the plants to continue growing. Students can return to make observations, both quantitative and qualitative, as the plants continue through their life cycle. Additional points to observe can be height of the plant and how long it took to break through the soil."

6. Students can create their own colorful illustrations or sketches of the flowers along with their observations and post them on a wall to create their own display, similar to Ehlert's pictures from the book.

Grades 4–6: From Seed to Flower

Purpose
Students will dissect a flower to investigate its structures.

Materials
- Flowers such as lilies, daffodils, iris, or other flowers with large parts
- Magnifying lenses
- Safety scissors or plastic knives to dissect plants
- Seeds identified in the book for inspection (a variety of different types and shapes/sizes)

Some plants are poisonous or have saps that can cause skin irritation, so have students wear gloves. Make sure students thoroughly wash hands after handling plants.

Procedure
1. Read *The Reason for a Flower* to the class, allowing students to focus on the pictures and overall story. Then, go back and reread the book to the class, asking students to focus on what is happening in the book. Ask questions such as "What types of organisms help move pollen from flower to flower?" and "What other ways can pollen be moved from flower to flower?" Continue to read the book, asking students to focus on the unfamiliar words (vocabulary associated with the parts of a flower).

2. Pass out flowers to individual students or small groups. After students have had a chance to examine the key parts of an actual flower, they should each select one flower and sketch it on a piece of paper or in their science notebooks. Ask students to identify the parts of their flowers and label their illustrations. After students examine the outer part of the flower, they can begin to dissect it. Have them try to find the following parts:

- *Stamen* (the male part of the flower; there are usually multiple stamens on each flower)
- *Filament* (the long stalk of the stamen)
- *Anther* (located at top of the filament; holds pollen)
- *Pistil* (the female part of the flower)
- *Stigma* (located at the top of the pistil; collects pollen)
- *Style* (carries pollen from the stigma, down through its hollow body, to the *ovary*, where the pollen fertilizes the flower's eggs)

3. Students may ask about which flowers are male and which flowers are female. Explain that some flowers have all-male or all-female parts (melons and pumpkins for example), but most flowers are considered *perfect,* in that they have both male and female parts. Why might perfect flowers want to attract pollinators like bees? (to ensure *cross-pollination*—pollen from different plants). Do they see any structures on the flower that might attract pollinators? (bright colors, nectaries).

4. Students can continue to sketch and label the parts of the flower as they continue to dissect them using both their eyes and a magnifying lens to get a close-up look at the parts. As students focus in, they will be able to see specific features and possibly tiny eggs or ovules in the pistil's ovary (at the bottom of the style).

5. After students have had an opportunity to inspect the parts of a flower, have them do the same with seeds. Thus far, this activity has focused on how flowering plants reproduce. The outcome of this reproduction is seeds, which come in many different sizes and shapes. By having students examine the different parts of a seed, they will have a better idea of where their flowers came from initially.

A follow-up to this activity could be comparing and contrasting flowering and nonflowering plants, such as mosses, and exploring their methods of reproduction.

Reference

National Research Council (NRC). 1996. *National science education standards.* Washington, DC: National Academies Press.

Chapter 27

Crazy for Loco Beans

By Karen Ansberry and Emily Morgan

Do you remember your amazement the first time you watched a Mexican jumping bean hop around in the palm of your hand? Investigating "jumping beans" is a unique way to get your students engaged in scientific inquiry. Here, students investigate mystery objects (Mexican jumping bean seed pods) using a multicultural story book and the 5E Instructional Model.

Trade Books

Lucas and His Loco Beans: A Bilingual Tale of the Mexican Jumping Bean

By Ramona Moreno Winner, illustrated by Nicole Valesquez (with flip book by Mary McConnell)
Brainstorm 3000, 2002
ISBN 978-0-9651174-1-8
Grades K–4

SYNOPSIS

Bilingual rhyming text tells the tale of a young boy who learns about the life cycle of the Mexican jumping bean moth from his grandfather. The story is told primarily in English, interspersed with Spanish words and phrases. Note: On pages 15 and 26, the jumping bean moth larva is incorrectly referred to as a "worm." Replace the word worm *with* larva *as you read the book aloud.*

A Monarch Butterfly's Life

By John Himmelman
Children's Press, 2000
ISBN 978-0-516-26537-7
Grades K–4

SYNOPSIS

Concise text and realistic color paintings highlight the migration of a monarch butterfly as she journeys south to Mexico and back again. Simple sentences briefly describe each stage of development in the monarch's life cycle, and italicized words are defined in the glossary.

Curricular Connections

In the activities that follow, *Lucas and His Loco Beans* is read aloud to lead students through a scientific inquiry. We suggest rereading the story at a later time with the purpose of exploring the Spanish language. Students who speak English only can use the story and illustrations to infer the meanings of the Spanish words. This book also provides an opportunity for students who speak Spanish to share their language skills with their classmates and can be a springboard to learning more about Hispanic cultures. The use of bilingual books in the classroom suggests that teachers value other languages and cultures and provides an effective tool for raising all students' awareness of diversity by exposing them to different languages (Ernst-Slavit and Mulhern 2003).

The accompanying activities touch on *National Science Education Standards* scientific inquiry and life science standards, as K–3 students explore "mystery objects" and learn about the life cycle of the jumping bean moth and later compare its life cycle to that of another insect found in Mexico, the monarch butterfly (NRC 1996). In the grades 4–6 student activity, students experiment to see if temperature affects the rate at which the beans "jump."

Mexican jumping beans are actually sections of the seed pods of a Mexican shrub that contain the larvae of a small moth. For a list of vendors, see Internet Resources. The USDA permits importation of jumping beans into the United States because the moths cannot infest local plant species.

If kept at a cool temperature (above freezing), lightly misted with water each week and then dried with a paper towel to prevent mold, jumping bean seed pods can be easily maintained in the classroom for several months. These activities are best done in late summer or fall when active seed pods are available. Jumping beans are a choking hazard for children under four.

Grades K–3: Loco Bean O–W–L

Materials

- Jumping beans (one per student)
- My Mystery Object O-W-L chart (p. 151; one per student)
- Chart paper
- Markers

Engage

Place a jumping bean seed pod in each student's palm and ask them to sit very still while watching the object for a few minutes. Do not tell them what the objects are. After a few minutes, the heat from the children's hands will cause the larvae and seed pods to move.

Explore

Give each student a copy of the O-W-L (Observations-Wonderings-Learnings) chart, a hand lens, a ruler, and one seed pod. Have students draw the "mystery objects" and then record observations and measurements. Next, have them list their "wonderings" about the objects. Ask students to share some of their wonderings with the class, and allow them to share their inferences about the identity of the objects. Some students may already be familiar with Mexican jumping beans and will be eager to "spill the beans"!

Explain

Tell students that you have a book to share that may help them solve the mystery. Use paper to hide the cover of the book *Lucas and His Loco Beans* and don't tell students the title of the book! Explain that you want them to make some guesses about what the objects are without seeing the cover or title yet. Older students can write down their inferences about the mystery objects as they get clues from the story. Read the book aloud, making sure to replace the word *worm* with *larva*. When finished reading, share the cover and the title. Have students explain

Crazy for Loco Beans

to a partner what they learned about the objects. From the reading, students will discover that "loco beans" are actually seed pods that contain moth larvae. Explain that the set of stages an animal goes through in its life is called a *life cycle*. Show students the diagrams on pages 26–29 and talk about each stage in the life cycle of the jumping bean moth. Have students add their learnings from the reading to the "L" column of their O-W-L charts.

Elaborate

Ask students if they can describe the life cycles of other insects, and allow time for sharing. Show them the cover of the book *A Monarch Butterfly's Life*. Ask them to think about how the life cycle of a monarch butterfly compares to the life cycle of a jumping bean moth as you read the book aloud. After reading, make a Venn diagram on chart paper and discuss similarities and differences of the two life cycles (see Figure 27.1).

Evaluate

Give students pictures of monarch butterfly and Mexican jumping bean moth life cycle phases and have them put them in the correct order, or have students create their own labeled drawings in the correct order. Ask students to describe each stage either orally or in written form.

Grades 4–6: "Loco–Motion"

Materials

- Jumping beans (one per student)
- The Mexican Jumping Bean article (p. 152)
- Poster paper
- Markers

Engage

Give each student a jumping bean seed pod and have them observe its movements for a few minutes. Tell

Figure 27.1. Life Cycle Venn Diagram

Jumping bean moth

Lays eggs on flowers of a shrub

Eats inside of seed pod

Wraps up in a silk cocoon

Stays in Mexico its entire life

Both

Hatch from eggs

Live on plants

Grow wings and fly

Live some time in Mexico

Monarch butterfly

Lays eggs on milkweed plant

Eats milkweed leaves

Skin hardens and becomes a chrysalis

Migrates north to the United States

students that you have a book to share that may help them determine what makes the objects move, then begin reading aloud *Lucas and His Loco Beans*. From the reading, students will discover that "loco beans" are actually seed pods that contain moth larvae. Brainstorm a list of variables that might affect the movement of the seed pods (e.g., light, heat, sound, and vibration). Next, reread page 19, "Your warm body makes them move like crazy," and ask them to brainstorm ways they could test the effect of heat on jumping bean movement.

Explore

Have teams of students design experiments to answer the question, "Do jumping beans move more at warmer temperatures?" Be sure students know that excessive heat or freezing will kill the larvae. One way to test the question would be to count the number of movements on the table-top in one minute and compare that to the number of movements in the palm of their hands in one minute. Make sure they hold their hands as still as possible. Movement can make jumping bean larvae go still. Encourage students to control all other variables, to repeat their experiment multiple times, and to organize their data in a table and/or graph. You may want to have students design further experiments on jumping bean movement.

Explain

Have teams share their conclusions using data as evidence to support their explanations. Students should find that jumping bean seed pods move more at warmer temperatures. Ask students to think about some possible reasons for this phenomenon.

Elaborate

Have students read the article titled "The Mexican Jumping Bean" to find out what causes the jumping bean larvae to become more active at warmer temperatures. The article also describes the fascinating life cycle of the jumping bean moth.

Evaluate

Teams of students can create posters to share what they learned about Mexican jumping beans. The posters may include data and other information from their experiments, labeled diagrams of the life cycle of the Mexican jumping bean moth, further questions, and any other information they feel is important to include.

References

Ernst-Slavit, G., and M. Mulhern. 2003. Bilingual books: Promoting literacy and biliteracy in the second-language and mainstream classroom. *Reading Online*, 7(2). *www.readingonline.org/articles/art_index.asp?HREF=ernst-slavit/index.html*

National Research Council (NRC). 1996. *National science education standards*. Washington, DC: National Academies Press.

Internet Resources

Jumping Bean Vendors

Jumping Beanditos
   www.jbean.com

JumpingBeansRUs.com
   www.jumpingbeansrus.com

My Pet Beans
   www.mypetbeans.com

Wayne's Word: An On-line Textbook of Natural History
   http://waynesword.palomar.edu/plaug97.htm

Crazy for Loco Beans

Name: _____

My Mystery Object

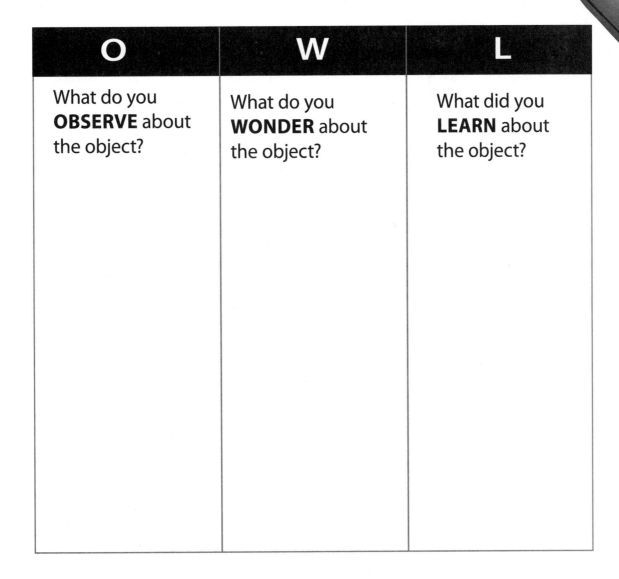

O	W	L
What do you **OBSERVE** about the object?	What do you **WONDER** about the object?	What did you **LEARN** about the object?

The Mexican Jumping Bean

What Is a Jumping Bean?

A Mexican jumping bean is not really a bean. It is part of the **seed pod** of a shrub found in Mexico. Inside the seed pod lives the **larva**, or caterpillar, of a small moth. The larva finds everything it needs to stay alive inside the seed pod: food, water, air, and shelter. It gets food by eating the inside of the seed pod. It gets water and air that enter the thin shell of the seed pod. The seed pod also gives the larva protection from weather and predators.

How Does It Get Inside?

In the early summer, a female moth lays her eggs on the flowers of a certain shrub found in the mountains. No other kind of plant will do. When the eggs hatch, the tiny white caterpillars chew their way into the seed pods of the shrub. The seed pods fall to the ground and split into three sections. Each section can contain a single larva. Not all of the sections contain larvae. If they did, there would be no seeds left to grow into Mexican jumping bean plants!

What Makes It Jump?

The larva spins silk to line the inside of its new home. When it gets too hot, it grabs onto the silk and snaps its body to make the seed pod jump and roll. When the larva reaches a cooler spot in the shade, it doesn't move as much. If you could hear the seed pods moving around on the dry ground, you might think they sound like rain drops!

How Does It Become a Moth?

The larva spends the summer and fall eating and moving around inside its seed pod. It chews a tiny flap in the seed pod called an exit hole. As winter approaches, it begins to spin more silk until it is covered with a soft cocoon. The larva begins to go through many changes. It is now called a pupa. The pupa slowly turns into an adult moth. These changes are known as metamorphosis. The following summer, the adult moth hatches out of the cocoon and squeezes through the exit hole it made months earlier. Soon it will find a mate, lay eggs, die, and the cycle will begin again.

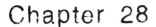

Chapter 28

Seeing and Sorting Seashells

By Christine Anne Royce

Students often become a bit wistful for summer once the school year starts. Why not initiate a seashell classification activity to re-spark their interest in science and learning and to capture the excitement of the summer! Children love to handle and examine shells at any time of the year—with these favorite nature objects, you often can't get kids to stop talking and sharing their observations about them! Whether working with commercially purchased shells or treasured collections of your own, students will delight in seeing how each shell is unique but can be connected to many others through classifications based on its characteristics.

Trade Books

Seashells by the Seashore
By Marianne Berkes, illustrated by Robert Noreika
Dawn Publications, 2002
ISBN 978-1-58469-034-4
Grades K–3

SYNOPSIS
Rhyming text tells the story of a young girl collecting a dozen different shells as a gift for her Grandma's birthday. In addition to illustrations and a brief description in the text, detailed information about each shell is provided in a pull-out guide at the end of the book.

Seashells, Crabs, and Sea Stars
By Christiane Kump Tibbitts, illustrated by Linda Garrow
Northword, 1991
ISBN 978-1-55971-542-3
Grades 2–5

SYNOPSIS
Seashells, Crabs, and Sea Stars *covers those three categories with colorful drawings and information about each object, including where each can be found. The book also includes craft projects at the end of each section that may provide an art integration.*

Curricular Connections

Anyone who has had the opportunity to visit a coast has probably picked up a seashell—either a fully formed shell that has survived the trip to the beach or at least pieces of broken shells that have met their match with the forces of nature. Either way, the individual patterns, formation, and beauty of these shells are evident. Seashells are made by mollusks. Shell-making mollusks are born with tiny shells that grow throughout their lives. Many students have the misconception that a shell is a mollusk home. It is more accurate to think of a shell as part of a mollusk's body.

Using seashells in the classroom provides the teacher with an opportunity to connect process skills such as classification, life science concepts associated with characteristics and adaptations of living things, and science as inquiry through the asking of investigable questions.

Many of the shells described in *Seashells by the Seashore* have characteristics that lend themselves to the name of the shell—for example, a scallop shell has a scalloped edge and a kitten's paw has the rough shape of a cat's foot. In making observations of shells, younger students tend to describe the properties of the object including shape, color, and size, helping students explore properties of objects. This can be a beginning point and students can advance to using rulers, balances, or other measuring tools.

Students in the older grades take the observation of characteristics a step further by classifying shells into groups—initially of their own choosing and then by classifying the shells as *univalve* or *bivalve*—a key element in classification due to the structure of the shell. Although this activity focuses on the idea of classifying shells, students can take their observations about shells and begin to develop questions they have about either the shells or perhaps the animals that once made them.

Grades K–3: Seeing Seashells

Purpose

Make observations about different types of seashells and then describe a single shell using observable characteristics.

Materials

- A collection of shells for each group (shells mentioned in the book include periwinkles, kittens' paws, jingle shells, olive shells, scallops, whelks, oyster shells, slipper shells, moon shells, pen shells, ark shells, and cockle shells)
- Copies of illustrations of the shells to refer to throughout the activity (*www.seashells.org/ identifying.html* is a good site for pictures and descriptions of shell types)

Note: Shells for use in this activity are usually available for purchase at different types of craft stores or could be ordered from a science supply house.

Procedure

1. To engage students, ask, "Have you ever been to the seashore and picked up seashells?" or, "Does anyone have a collection of seashells?"

2. Begin by reading *Seashells by the Seashore* to the class. The rhyming pattern of the book also incorporates counting, which will appeal to younger students. Remind students to focus on the illustrations and the different names of the shells. As each type of shell is introduced, post its illustration for students to refer to. Some (but not all) of the shell names give a clue to identification of the shell.

3. After reading the story, return to the shell illustrations. Ask students to make observa-

Seeing and Sorting Seashells

tions, and write the different observations or descriptive words next to each shell picture. Some of the descriptive words might be *speckled* or *solid* for the color or *smooth* or *ridged* for the feel of the shell.

4. After students have made some initial observations of the shells found in the book, provide each group of students with a collection of shells. Ask each student group to spend time making observations of the shells. They should consider how these "real" shells are similar to and different from the pictures they looked at. This helps students realize that the same type of shell may have similarities but not the same exact look—for example, the color or the size are characteristics that are within a certain range for each type of shell but have some variation. Possible answers may include, "The pictures don't show as much detail as the real shells" or, "The real shells and the pictures don't look the same." This is a good place to introduce the idea that not only do different types of shells look different, but the same type of shells can also look different from each other.

5. Ask each student to select a single shell from the pile and write a description of the shell. They should also draw pictures of their shells, including as many details as possible. Older students can also make measurements of their shells.

6. After students have had an opportunity to describe their shells, they can individually share their descriptions with the class. Students can then see if they can match the descriptions being read with the shells at their stations.

7. By using observations and descriptive writing, students are connecting the science-process skills to another area that is important in science—reporting their findings and information to others.

Grades 4–6: Sorting Shells

Purpose

Develop a classification system for shells and classify them as a univalve or bivalve.

Materials

- A collection of shells for each group (suggested shells include periwinkles, kitten's paw, jingle shells, olive shells, scallops, whelks, oyster shells, slipper shells, moon shells, pen shells, ark shells, and cockle shells)
- Poster paper
- Markers

Procedure

1. Either read the section on seashells from *Seashells, Crabs, and Sea Stars* to the class or provide a copy of the book to each work group to use as a reference.

2. Distribute a set of shells to each group, and ask the students to examine the shells carefully and record their observations of the shell size, color, shape, and texture.

3. After students make observations, ask them to sort the shells into some system based on the properties they observed. The teacher can then ask students to share their shell observations with the whole class, then lead a discussion on the idea of classification and how we classify—by grouping objects together through some type of similar attribute. Have the groups return to their particular collection of shells and discuss ways they could classify their shells and develop a classification system. They should be able to explain the attributes they chose and the methods by which they grouped their shells. Some groups may choose size, others may choose color, and others may choose shape.

4. Ask the class, "How do you think scientists group the different types of shells?" The pos-

sible answers may include where it was found and what animal made it. At this point, the teacher should introduce the vocabulary words *univalve* and *bivalve*. A univalve is a single shell that contained an animal whereas a bivalve has two shells that are symmetrical along a hinge line (usually you will only find a single part of the shell when collecting seashells). Univalves and bivalves move and eat in different ways—it is one of the methods of classifications for shells.

5. Ask the students to reexamine the shells they have and then sort them into groups according to this classification—univalve or bivalve. Some students may have a difficult time with the idea that the bivalves only have "one shell" in the examples they have. The teacher can discuss with the class what might have happened to the other part of the shell. It will allow students to understand why we don't find bivalves in full pieces when collecting seashells.

6. Once students have sorted the shells into groups based on their "valve" status, ask the students to

make some observations about the structure of the shell: "Where would the animal have been attached in each type of shell? How do you think the animal moved and obtained food?" Univalve shells have a space for the animal to retreat inside, whereas bivalves can close their shells to protect themselves.

7. Students can then begin to develop investigable questions around their new knowledge of shells. Questions such as "What are shells made of?" might be answered through additional research. Questions such as "Which shell has the most volume inside?" might be answered through an experiment based on the properties of a shell.

Reference

National Research Council (NRC). 1996. *National science education standards.* Washington, DC: National Academies Press.

Chapter 29

Unusual Creatures

By Karen Ansberry and Emily Morgan

Triops, also called *tadpole shrimp*, are tiny aquatic animals that are easy to raise and fun to watch. Their rapid growth rate, underwater acrobatics, and entertaining feeding behaviors make these inexpensive and readily available animals well-suited for classroom observation. In these activities, students observe "mystery eggs" that quickly hatch and grow into triops larvae. Students learn to use a science notebook to keep track of their observations, wonderings, and learnings about the fascinating characteristics and life cycles of these most unusual creatures.

Trade Books

Triops: A Very Unusual Creature

By Helen Pashley and Lori Adams.
Little Science Books, 2009
ISBN 978-0-9824412-0-6
Grades K–6

SYNOPSIS

Triops reveals the unusual looks and amazing survival skills of this fascinating crustacean with stunning close-up photographs of this remarkable animal.

Crab Moon

Written by Ruth Horowitz, illustrated by Kate Kiesler
Candlewick Press, 2004
ISBN 978-0-7636-2313-5
Grades 3–6

SYNOPSIS

This NSTA/Children's Book Council Outstanding Science Trade Book tells the story of a young boy who watches horseshoe crabs come ashore to lay their eggs. The end includes information about the life cycle of this ancient creature.

Curricular Connections

Triops are small, freshwater crustaceans that have a relatively short lifespan and grow quickly to their adult length. They are similar in appearance to horseshoe crabs, although the horseshoe crab is not a true crab or even a crustacean. However, due to the similarities in their appearance and the development of their larvae, scientists think that triops may be related to the horseshoe crab. Comparing these two creatures can help students fine-tune their observational skills and encourage them to ask questions and learn more about each of these ancient animals. The *National Science Education Standards* suggest that students in grades K–4 should be provided with opportunities in which they can ask questions about objects, organisms, and events in the environment and use their observations and reliable resources to formulate explanations (NRC 1996). In grades 5–8, the Standards suggest that students understand that different kinds of questions require different kinds of investigations. Investigations involve observing and describing objects, organisms, and events; seeking more information; discovery of new objects and phenomena; or making models. The lessons here involve investigations in which students must observe and describe and then seek information from reliable sources.

Grades K–2: Mystery Eggs

Materials

- Container of dried triops eggs
- Small aquarium
- Spring water
- Thermometer
- Poster paper
- My Animal Observation Notebook (pp. 161–162; one per student)

Engage

Follow the directions for setting up a small triops habitat in your classroom, using a small aquarium or other stable, transparent container (see Internet Resources). Place an aquarium thermometer in the tank or attach an adhesive thermometer strip to the outside of the tank. After you have set up the tank, gather students around it, and add a container of dried triops eggs. Tell students that you just added some "mystery eggs" to the water. Ask, "What kinds of animals hatch from eggs?" Have students predict what type of animal might hatch.

Explore

Create a large version of the O-W-L science notebook (pp. 161–162) using a folded sheet of butcher paper or large poster paper. Then create tabs to label the three sections: Observations, Wonderings, and Learnings. Discuss the kinds of things that a scientist studying the eggs might include in each section of the notebook. For example, in the observations section, a scientist might record the date, time, and water temperature. Remind students that they will not be tasting or touching the eggs! Observe the growing triops as a class for about two weeks. You may need to separate the triops into different containers (e.g., plastic salad containers) so that several small groups of students can observe them at the same time. This is also a good idea because triops tend to be cannibalistic! Provide hand lenses so students can get a closer look and record their observations and wonderings about the "mystery eggs" in their notebooks as they share them with the class.

Explain

Read aloud the wonderings page of the class notebook. Then show students the cover of *Triops: A Very Unusual Creature*. Tell them that the animals they have been observing are called *triops* and that you have a book that might answer some of their questions about these unusual animals. Read the introduction on page 3. Then model how to use the table of contents to find answers to specific questions on the wonderings page by choosing a question and reading through the titles listed. For example, the question "Where do these animals live?" could be answered by reading the section titled "Where in the World

Unusual Creatures

Do You Find Triops?" It is important for students to understand that when you pick up a nonfiction book, you don't have to read it from start to finish like a story. You can enter the text at any point. The table of contents is a handy tool for finding the part of the book that has the answer to your question. Also model how to use the glossary to find the meaning of bold print words, the labeled diagrams to find out the names of body parts, and the index to find the page where a specific word is mentioned. As a class, add some answers to the students' wonderings and other new knowledge about triops to the learnings section of the O-W-L notebook.

Elaborate

Give each student a copy of an O-W-L notebook (pp. 161–162). Provide students opportunities to practice their science notebooking skills by allowing them to observe other classroom pets, or even animals in the local ecosystem, for several days. You may want to send the notebooks home and have students record data about their own pets or friends' pets.

Evaluate

Have students share their completed notebooks with you. Check that they have their observations dated and ask them to share some of their observations, wonderings, and learnings with you. As they share their learnings, ask them how they found out.

Grades 3–6: Triops and Their Relatives

Materials (per group of three to four students)

- Container of dried triops eggs
- Small aquarium
- Spring water
- Thermometer
- Notebook
- Sticky notes

Engage

Follow the directions for setting up a small triops habitat for each group of three to four students. Tell students that you have some "mystery eggs" for them to put in their tanks and that you will give them some clues over the next several days to help them determine what might hatch from the mystery eggs. Give each group a copy of the care instructions that come with the eggs (see Internet Resources), but cover up the word *triops* wherever it appears in the instructions.

Explore

Have each student bring in a small notebook. Using sticky notes, have them create tabs for three sections in the notebook called Observations, Wonderings, and Learnings. Discuss the kinds of things that a zoologist (a scientist who studies animals) might include in each section of the notebook. For example, in the observations section, zoologists might record drawings and descriptions of what they see, hear, or smell. Remind students that they will not be tasting or touching the eggs! Have students make daily observations of their tanks, using their notebooks to record their data and questions. Periodically, ask students to share some of their observations and wonderings with the whole class.

Explain

After several days, ask each group to share their wonderings with one another and then decide on their most compelling question. Have them write that question on a large sticky note or sentence strip and post these on the board. Then, show them the cover of *Triops: A Very Unusual Creature*. Read the introduction on page 3 and the table of contents on the opposite page. Ask students whether there are any sections that might help them answer the wonderings on the board. As you read aloud the sections students choose, have them add to the learnings section of their notebooks. Ask students to continue recording their questions in the wonderings section of their notebooks as you read. Be

sure to read the section titled "Interesting Facts" on page 23 which explains *diapause*, the temporary stoppage of life processes, which can help triops eggs survive extreme conditions.

Elaborate

Read page 21, titled "Similar Creatures," that explains that triops may be related to the horseshoe crab. Ask students whether they have ever seen or heard of a horseshoe crab. Show them the cover of the book *Crab Moon*. Tell students that as you read the book aloud, you would like them to think about what triops and horseshoe crabs have in common and what is different about the two. After reading the book aloud, ask students to create a Venn diagram comparing and contrasting triops and horseshoe crabs in the learnings section of their notebooks.

Evaluate

Have students use the information they have collected in their notebooks to create a poster about triops. The poster might include a detailed, labeled drawing of a triops, a description of their life cycle (including an explanation of diapause), fascinating facts, similarities to and differences from horseshoe crabs, and a question they still have about triops.

Reference

National Research Council (NRC). 1996. *National science education standards.* Washington, DC: National Academies Press.

Internet Resources

Little Science Books
www.littlesciencebooks.com
My Triops
www.myTriops.com
Triassic Triops
www.Triops.com

Unusual Creatures

My Learnings

My Animal
Observation
Notebook

Zoologist's name:

Observations: from

to

My Wonderings

My Observations

DATE	DATE	DATE	DATE

Chapter 30

Can You See Me Now?

By Christine Anne Royce

"You can't see me" is a popular childhood taunt, and although young children may not understand that hiding in plain sight does not conceal them, it does connect with a lesson about camouflage. In these activities, students will learn that camouflage is essential to an animal's survival in nature and is accomplished through a variety of ways.

Trade Books

Red Eyes or Blue Feathers: A Book About Animal Colors

By Patricia M. Stockland, illustrated by Todd Ouren
Picture Window Books, 2005
ISBN 978-1-4048-0931-4
Grades K–4

SYNOPSIS

Stockland uses the question, "What's the best way to survive in the wild?" to introduce how an animal's coloration helps it survive in its native environments. The author discusses different animals, ranging from a rhinoceros to a polar bear to a macaw, so that young readers will have an opportunity to explore how an animal's color assists it in survival.

Hide and Seek: Nature's Best Vanishing Acts

By Andrea Helman, illustrated by Gavriel Jecan
Walker and Company, 2008
ISBN 978-0-8027-9690-5
Grades 3–6

SYNOPSIS

Hide and Seek introduces readers to different habitats around the world and the animals that live there. Through actual photographs, students can see the many ways that animals can be camouflaged and how the different locations impact their coloration.

Curricular Connections

Hide-and-seek is a common childhood game. From the youngest age, students think that if they hide well, they won't be found and become "it." As students become more skilled at this game, they quickly realize that blending in with their surroundings will make them more difficult to spot; thus, they have learned the basic concept of camouflage even if they don't realize it. This chapter focuses on the concepts of animal coloration and camouflage. These concepts are part of the *National Science Education Standards* (NRC 1996) in the area of life science. Specifically, younger students should understand that an "organism's patterns of behavior are related to the nature of that organism's environment, including … the physical characteristics of the environment" (NRC 1996, p. 129).

Students will be introduced to the idea that "an organism's behavior evolves through adaptation to its environment" (NRC 1996, p. 157). This idea connects to the story about the peppered moth in England that is presented in the activity for grades four through six. Although natural selection and adaptation are long-term features in evolution of a particular species, an organism's coloration assists it in avoiding predators and helping it survive in its daily environment through camouflage.

Grades K–3: In Plain Sight

Purpose

To observe the characteristics that help animals stay camouflaged in an environment

Materials

- Pictures of camouflaged animals from magazines or websites (see Internet Resources)
- Craft sticks
- Cones or flags for marking an outside area

Procedure:

1. Read *Red Eyes or Blue Feathers: A Book About Animal Colors* to the students. Stop at each two-page spread and ask students to make observations about what color each animal is and what color their environment is before reading the text (e.g., the fox is a rusty brown and lives in the woods, where there are brown tree trunks and leaves). While you read, have students focus on what helps the animals be camouflaged in their environment. Engage the class in a discussion about why the features help camouflage the animals.

2. Ask students to view pictures of animals and describe what type of location each animal should live in to blend in with that environment. For example, a picture of a squirrel might elicit responses such as "in a tree," but the teacher should ask, "Why does a tree help camouflage a squirrel?" Some students may suggest locations that a particular animal would not live in, which can be corrected, but remember that animal habitats are not the main focus.

3. Once students have described the type of environment that would be best for camouflage, give each student a picture of a different animal. Each picture should be on card stock or laminated and have a craft stick attached to the bottom. With the help of a coteacher, parent volunteer, or classroom aide, split the class in half and have the first team of students go outside to a predetermined area set off by cones or flags and "hide" their pictures in an area that camouflages the animals (e.g., a picture of a chipmunk could be put into the ground next to a rock or under a bush). The pictures should not be placed flat on the ground; students should be able to see them from a standing position. The craft sticks are a great way to hide the picture upright.

Can You See Me Now?

4. Once the first team has placed their pictures, the other half of the class should come to the location and be the predators who are hunting for the prey (the pictures). Repeat this process by having the teams switch places.

5. Once students have had an opportunity to be the predator (looking for the pictures), and the prey (those that hide the pictures), bring them together to discuss questions such as, "Was it hard to find a place to 'hide' the picture? Why or why not?" Some pictures should be brightly colored—such as the brilliant macaw in the story—and will not be well camouflaged in the setting near your school. You can focus on other questions such as, "What types of animals were well camouflaged in this area?" or "If the animal was moving would it be easier or harder to find them?" Return to the story to discuss why some animal adaptations have evolved over time based on their environments.

Grades 4–6: Moth Madness

Purpose

To examine camouflaging in an environment by hunting for moths in the classroom

Materials

- Various types of paper and cloth
- 3 to 4 in. moth pattern (p. 167)
- Tape
- Locating Moths Data Table for students (p. 168)

Procedure

1. Prior to class, using the moth pattern, cut out two to three moths per student in different colors and textures. You can use material such as construction paper, newspaper, or felt to ensure that you have a collection of different colorations. Hide moths around the classroom in obvious places; make many of them blend into their environment (e.g., a blue moth on a blue poster) and place others where

they do not blend into their environment (e.g., a black moth on a white wall). Make a mental note of where you hid the moths.

2. Introduce *Hide and Seek* to the students by reading them the entire book or the sections dealing with camouflage associated with coloration. Focus on the different types of animals described in the text and where they are hiding in the illustrations. Use the text and pictures to ask students to define *coloration* and *camouflage*.

3. Explain to students that many animals hide and survive in their natural environments because they are camouflaged in their surroundings. Share the story of the peppered moth in England during the 1800s (see Internet Resource).

4. Have each student create (or provide them with) a data table (p. 168). Explain the different components of the table so that students are aware of the type of information they will record.

5. Generate ideas from the students about how they will locate the moths. Possible answers might be, "Look for colored moths on similar colored backgrounds" or, "Look from different angles by moving your head." Animals do this when they locate a prey in hiding, partly through adaptations in their eyesight (e.g., ability to see in nocturnal conditions) and partly by moving their head back and forth to gain a better vantage point.

6. Show students a sample moth and explain that you have placed moths around the room. All moths are in view (i.e., not hidden in drawers or cabinets). While sitting in their seats, provide the students with three minutes to find and record as many months as they can. Students should look quietly by moving around the room for three additional minutes so they do not give away the location of the moths to other students; tell them that their own survival as a predator depends on this.

7. Have students return to their tables and decide which moths were the easiest to locate and which were the most difficult to locate. Locate all of the moths originally placed in the room. There may be some moths that the students did not find.

8. Conclude with a discussion of coloration and camouflage. Discuss what made these particular moths so difficult to find. Use guiding questions such as, "What features make some moths more visible than others?" and "What are some colors or patterns that would help moths survive in a meadow or woodland environment?"

Reference

National Research Council (NRC). 1996. *National science education standards.* Washington, DC: National Academies Press.

Internet Resources

Animal Camouflage Pictures
http://animals.howstuffworks.com/animal-facts/ animal-camouflage-pictures.htm

Nova Online Seeing Through Camouflage
www.pbs.org/wgbh/nova/leopards/seeing.html

The Peppered Moth
www.millerandlevine.com/km/evol/Moths/moths. html

Moth Pattern

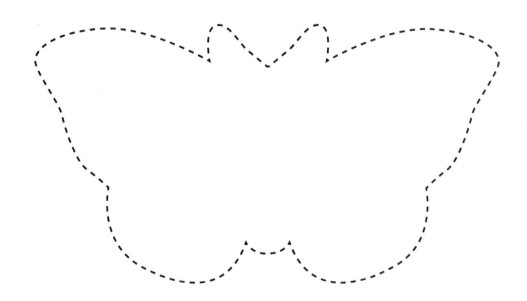

Locating Moths Data Table

#	LOCATION OF MOTH	WELL CAMOUFLAGED? (YES OR NO)	WHY OR WHY NOT?	SEATED OR MOVING AROUND ROOM?
1	On window shade	Yes	Tan colored moth on tan shade	Seated

Chapter 31

Survival Skills

By Christine Anne Royce

Animals have many adaptations that help them survive in their environments and meet their special needs of food and shelter. Two such adaptations are camouflage and beaks.

Trade Books

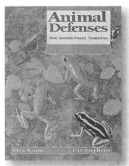

Animal Defenses: How Animals Protect Themselves

By Etta Kaner, illustrated by Pat Stephens
Kids Can Press, 1999
ISBN 978-1-55074-419-4
Grades K–8

SYNOPSIS

Animal Defenses *explores different defense mechanisms animals use to protect and defend themselves, such as camouflage, trickery, and warning colors. Each section discusses various animals—including some unusual examples—that use one of the approaches and explains the strategy with accessible text and colorful illustrations.*

Beaks!

By Sneed B. Collard III, illustrated by Robin Brickman
Charlesbridge, 2002
ISBN 978-1-57091-387-7
Grades K–4

SYNOPSIS

This beautifully illustrated book begins by discussing what birds do not have and quickly arrives at the point that birds have beaks, and beaks are enough. Chapters on different types of birds describe movement, food choices, and adaptations. The end poses a collage of birds and asks the student to test their "beak-ability" in identifying the type of food the bird would eat.

Curricular Connections

Both books this month are informational in nature, and the activities demonstrate an important use of trade books in the science classroom—that of gathering information. This skill is important at all grade levels and connects to many subject area standards. *Animal Defenses* poses the initial question to students, "What do you do when you are afraid?" Throughout this book, the reader learns of ways animals protect and defend themselves from predators. The book works well as an introduction to a study of animal defenses, or it can be used periodically as a reference throughout a unit on animals or adaptations.

In the activity presented here, young students examine the pictures that deal with camouflage and then read (or, if nonreaders, listen to the teacher read) the text to gather information about this topic. The first activity investigates a survival strategy that helps protect animals.

Older students may be more interested in exploring animals' structural adaptations that help them survive, based on their needs and environments. A bird's beak is just such a structural adaptation—some (like a sparrow's) are pointed and thick, which helps them crack open and eat seeds; others (like an eagle's) are curved and sharp—which helps in the tearing of their daily catch.

After reading *Beaks!* older students engage in a simulation in which they predict which food is best obtained by different beaks. After their exploration, students can refer to the book to confirm their observations and conclusions.

Grades K–3: Hide and Seek

Purpose

Students investigate how camouflage helps animals blend in with their surroundings.

Materials (per group of two to three students)

- Five different plastic animals, various colors
- Three different pieces of fabric, each about 15 cm × 15 cm and colorful—think floral prints
- Three butterflies cut from different-color construction paper, about 7 cm × 7 cm
- Clear tape
- Crayons, colored pencils, or markers (optional)

Procedure

1. Begin by asking the students to briefly examine *Animal Defenses* and identify some ways animals protect and defend themselves. When the students arrive at the possible answers of animals "hide" or "blend into the background," introduce the term *camouflage* if they are unfamiliar with it.

2. Focus students' attention on pages 10–15, which are devoted to camouflage. Ask students, "What helps animals blend in with their backgrounds? What do you notice about the different markings or colors on the animals and the backgrounds that they blend into best?

3. Ask each group of students to spread out the fabric pieces on a flat surface in front of them and make observations about the patterns and colors. Next, have each member of the group pick one plastic animal and note its colors. They should predict which fabric the animal would be best camouflaged against and explain their reasons.

4. After students make their predictions, ask them to place the animal against each piece of fabric and determine where it is best camouflaged.

5. After all students have "camouflaged" their animals, they can circulate around the room and try to locate the camouflaged animals of another group.

Now that students have had an opportunity to practice with the idea of camouflage, enlist a teacher's aid or parent volunteer for the second part of this activity:

Survival Skills

6. Distribute the paper butterflies to the students and allow them to decorate them or simply keep them a single color.

7. Divide the class in half and ask one group to step into the hallway (hence the need for a teacher's aide or volunteer) while the other group hides their butterflies so they are camouflaged. Suggestions include edges of bulletin boards, covers of books, posters, and so forth.

8. Once all butterflies are hidden, ask the remainder of the class to come in and find the butterflies. When students spot a butterfly, they should raise their hands. Select a student to describe the location of the butterfly to the class and retrieve the butterfly.

9. After all butterflies have been found, repeat steps 2 and 3 with the other half of the class. It will be amazing how much quicker it goes the second time around—although it must be pointed out that sometimes hidden butterflies are forgotten about and not found for months!

After participating in these activities, students can discuss where they have seen camouflaged animals in real life—at the zoo or a local park. The teacher can return to *Animal Defenses* at different points and engage the students in other activities that help them understand other defense mechanisms that animals use to survive or allow the students to investigate on their own.

Grades 4–6: Beaks Are for the Birds

Purpose

Students explore how a bird's beak helps it obtain food.

Materials *(per group of two to three students)*

- Two 5–7 cm deep plastic containers
- Plastic basin with water

- Sample "foods":
  * Marbles
  * Plastic beads (both small and large)
  * Toothpicks
  * Various nuts and dried beans
  * Styrofoam packing pieces (not the environmentally friendly kind that dissolve)
  * Rice
  * Rubber erasers
- One each of the following tools to represent beak types:
  * Tweezers
  * Pliers
  * Chopsticks
  * Spoons
  * Toy shovel
  * Eyedropper
  * Small strainer
  * Tongs
- Beaks Are for the Birds data sheet listing the "foods" across the top and the "beak types" down the left-hand side (p. 173)

Procedure

1. Start by asking students to predict what the birds on the front cover of *Beaks!* eat, describe the shape of each beak, and connect that shape to the predicted food.

2. Each member of the group should select one of the "bird beaks" to use and discuss what type of food they believe they will be most successful in obtaining and what type of food will be difficult due to the shape of their "beaks."

3. Each group should examine the items in each of its containers—small pieces in one; larger food pieces in another; and the rubber erasers, Stryofoam packing peanuts, and rice floating on the surface of the water basin. Then, students can discuss what food each object might represent (e.g., beads could be insects or seeds).

4. After discussion, students will predict how many pieces of food they will be able to pick

up with their beaks and record it. It should be noted that the students will be trying to get as much food as they can in a certain time period; however, birds are selective in what they eat and do not waste their time and energy trying to obtain food they don't eat.

5. Set a timer of one minute and ask students to pick up as many pieces of food from their basin as they can with their beaks. They cannot use their hands for anything except to steady the container.

6. Have students count how many pieces of each food type they were able to pick up and record the data.

7. Allow the students to switch "beaks" or food containers and repeat the time trial.

8. After students examine their predictions and actual counts, ask: "Was a particular beak type better suited to picking up different food types? If so, can you think of an example of a real bird and what they eat that would match this?" At this point the class can refer to *Beaks!* to see if they were accurate in their statements.

Reference

National Research Council (NRC). 1996. *National science education standards.* Washington, DC: National Academies Press.

Survival Skills

Beaks Are for the Birds

DATA SHEET

BEAK TYPE	FOOD SOURCE							
	Rice (Seeds)		Beads (Insects)		Nuts (Nuts/Seeds)		Erasers (Fish)	
	Predicted	Actual	Predicted	Actual	Predicted	Actual	Predicted	Actual
SMALL SPOON								
LARGE SPOON								
TWEEZERS								
SHOVEL								
PLIERS								
CHOPSTICKS								
EYEDROPPER								
STRAINER								
TONGS								

Chapter 32

Antarctic Adaptations

By Christine Anne Royce

When students think of cold weather, they often think of mittens, snowflakes, and penguins! Celebrate those tuxedo-clad birds that have come to symbolize winter and provide students with the opportunity to investigate adaptations that help penguins survive in their environment.

Trade Books

The Emperor's Egg
By Martin Jenkins, illustrated by Jane Chapman
Candlewick Press, 1999
ISBN 978-0-7636-0557-5
Grades K–4

SYNOPSIS
The Emperor's Egg *tells the story of the male emperor penguin. It describes how he cares for his mate's egg and eventually the young penguin chick while waiting for his mate to return from the sea. This wonderful book has illustrations of how penguins adapt to the cold Antarctic environment from moving around in huddles to playing by sliding on their stomachs.*

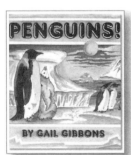

Penguins!
By Gail Gibbons
Holiday House, 1998
ISBN 978-0-8234-1388-1
Grades 2–5

SYNOPSIS
Gail Gibbons provides the older reader with information on the behavior, habitat, and physical characteristics of some of the 17 different types of penguins. Clearly illustrated diagrams help readers understand the text and do a nice job of introducing new vocabulary to the student.

Curricular Connections

Penguins are popular with young children, so naturally learning about them has found its way into many elementary curricula. Why are penguins so intriguing? Perhaps it is because they are cute and waddle around in social circles or because they are unlike most common birds children see. Beyond being cute, though, penguins provide an opportunity to investigate how different adaptations allow animals to survive in their environments, tying into *National Science Education Standards* for both younger and older elementary students.

The *National Science Education Standards* state that in grades K–4, students begin to "build understanding of biological concepts through direct experiences" (NRC 1996, p. 127). Although students can't engage in direct experiences with penguins, these role-play activities for students can help them understand penguins' and other organisms' characteristics and adaptations for survival. They also foster understanding that "each animal has different structures that serve different functions in growth, survival, and reproduction" (NRC 1996, p. 129).

Older students can investigate more in-depth questions about how penguins survive in their often frozen environment and develop geographic awareness. Gail Gibbons's *Penguins* begins by identifying where many of the different species of penguins live on a map, introducing students to the southern hemisphere and conditions in the Antarctic. More information about the life cycle and behavioral characteristics of penguins follows.

Typically, after reading the book, students ask such questions as, "How do penguins get food?" or "How do they stay warm?" Investigating questions such as these connects directly to the *National Science Education Standards* for life science for grades 5–8, which state, "biological adaptations include changes in structures, behaviors, or physiology that enhance survival and reproductive success in a particular environment" (NRC 1996, p. 158).

Grades K–3: Animal Adaptations

Purpose

Students simulate walking like penguins to better understand animal adaptations.

Materials

- Softball
- Oversized pillow (either a full-length body pillow or king-size pillow will do)
- Piece of rope or a scarf large enough (about 4 ft. long) to tie around the pillow and a student

Procedure

1. Before reading *The Emperor's Egg* to the class, engage students' prior knowledge by asking, "Have you ever seen a penguin? If so, where? Is that the penguin's natural environment?" This will help students realize that penguins they see in zoos or aquariums are real birds but not in their natural environment.

2. Begin reading the book to the students and have them make observations about penguins.

3. Next, discuss what the penguins eat, show the students a picture of an emperor penguin, and ask them to make observations about the beak, including about its shape. (Students may compare it with a clothespin or hook.) Ask, "How do you think a penguin gets its food?" At this point, introduce the word *adaptation* to the students and provide examples of adaptations—such as camouflage or flippers. While discussing adaptations, revisit the beak of the penguin and how it helps the penguin catch its food. Show pictures of other birds and their beaks. Good choices include sparrows (which have triangle-shaped beaks for cracking seeds) and hummingbirds (which have long and narrow beaks for sipping nectar from inside flowers).

Antarctic Adaptations

4. As you continue reading, students will discover that the male emperor penguin has the responsibility of caring for the egg until it hatches. Have students make observations about how the father penguin sits, moves, and cares for the egg.

5. Ask a student volunteer to remove his or her shoes and stand with feet about 4 to 6 cm apart (2 to 3 in.). Then place the softball on the indentation between the student's feet. The softball represents the egg. Referring to the illustrations in *The Emperor's Egg*, discuss how the male penguin keeps the egg tucked under a flap on the penguin's stomach, which is called a *brood pouch.*

6. Next, tell students they will each simulate the penguin's brood pouch by holding the pillow lengthwise against their stomachs so that the pillow reaches their feet and covers the egg. Once the pillow is in place, wrap the rope or scarf around the pillow and the student, and tie it in the middle of the chest. This serves as the penguin's large rounded stomach and brood pouch. Have the student waddle across the classroom (post a spotter in case a student loses his or her balance) while keeping the egg on top of his or her feet.

7. After all the students have tried this, ask them to react to the experience and describe the way they had to walk to keep the egg in place. Ask, "Was it easy to walk like a penguin?" "Can penguins move quickly?" Most students find it difficult to waddle around the classroom while keeping the softball on their feet, as it takes much coordination and prevents them from moving quickly. By "becoming penguins," students are able to imagine the different animal adaptations that are natural to each animal in its survival.

Grades 4–6: Cool Living Conditions

Purpose

Students experiment with how to keep their hands warm in ice-cold water to better understand penguin adaptations.

Materials

- Deep pan of ice-cold water
- Several gallon-size resealable plastic bags
- Enough solid shortening to fill one of the sealable plastic bags approximately three-fourths full
- Plastic gloves (the kind used for food preparation)
- Duct tape

Procedure

1. As shown in the opening pages of *Penguins!* there are 17 different types of penguins ranging in size from the little blue penguin to the emperor penguin. Gibbons goes on to show that all penguins live in the southern hemisphere, and although some penguins are found in New Zealand, South America, the Galapagos, and other places, the emperor penguin spends its entire life in the Antarctic. The temperature in the Antarctic can reach –60°C (–76°F). This varies with exact location and season, but regardless of where in the Antarctic a penguin may be, it needs to keep warm!

   Pose the following question to the students, "How do penguins stay warm in the Antarctic?" Allow students to brainstorm a list of ways. Some answers may relate to penguins having a thick skin or fur, which is incorrect (penguins actually have a very dense layer of feathers as their exterior coat). Other answers may focus on their fat or *blubber*; penguins actually have several adaptations that when combined, help keep them warm in this harsh environment.

2. Next, have one student in each group put on plastic gloves. The left hand will serve as a control or comparison against which to test the other options. Have the students place their right hands inside one of the sealable plastic bags and ask another student to wrap duct tape around the top of the bag near the student's wrist to seal the bag, trapping air inside. Teachers may want to model this step to show students how to tape the bag only and prevent students from taping skin. The plastic bag should have enough air in it to keep it inflated and away from the student's hand. Students then time how long they can leave each hand in the water before it gets too cold for comfort.

3. Allow the students to warm their hands before trying the next option. During this time, discuss the results from the previous test. The hand that was surrounded by air is similar to a penguin's feathers in that a penguin has an inside layer of down and an outer layer of feathers coated with a waxlike substance to help make the penguin waterproof. Air gets trapped between these two layers, and this helps insulate penguins from the cold air and water. Ask the students to describe ways we use trapped air to help insulate ourselves during the winter months. Examples include down-filled jackets and double-paned window glass.

4. Once the students' hands have warmed, they can try the next option—blubber! Repeat the process of the left hand serving as the control, but this time, have students insert their right hand (again wearing a glove also, to prevent a mess) into the sealable plastic bag full of shortening. Once their hand is in the bag, they should open and close their hands a few times to work the shortening around their hands. Students should submerge both hands again and time how long it takes for their hands to get too cold for comfort.

5. Ask students how the hand in the shortening compared to the unprotected hand or hand surrounded by air. Throughout this experiment, different students will have different time results for keeping their hand in the water due to their individual preferences and tolerance levels. However, students should realize that it is actually a combination of trapped air, waxy outer coating, and blubber that protect penguins from the extreme temperatures in which they live. These questions could begin a discussion about how penguins and other animals keep warm in the Antarctic.

Reference

National Research Council (NRC). 1996. *National science education standards.* Washington, DC: National Academies Press.

Chapter 33

The Mystery of Migration

By Emily Morgan and Karen Ansberry

The migration patterns of animals have long been a source of wonder and awe. From the 6-mile journey of the army ants in the rain forests of Costa Rica, to the 20,000-mile journey of the sperm whale through the world's oceans, these animals' ability to recognize when it's time to leave and where to go is a source of great fascination. The lessons described here explore what we know about animal migration and what still remains a mystery.

Trade Books

Going Home: The Mystery of Animal Migration

By Marianne Berkes, illustrated by Jennifer DiRubbio
Dawn Publications, 2010
ISBN 978-1-58469-127-3
Grades K–4

SYNOPSIS

This beautifully illustrated book uses rhyming verse to tell of the migration of 10 different animals. Insets and end matter contain specific information about the animals and what is known about their migration patterns.

Great Migrations: Whales, Wildebeests, Butterflies, Elephants, and Other Amazing Animals on the Move

By Elizabeth Carney
National Geographic Children's Books, 2010
ISBN 978-1-4263-0700-3
Grades 5–6

SYNOPSIS

Based on the National Geographic Channel's special Great Migrations, *this book features eight very different animals with one thing in common: They migrate.*

Curricular Connections

The *National Science Education Standards* (NRC 1996) suggest that in grades K–4, students should learn that an animal's patterns of behavior are related to that animal's environment and that when the environment changes, some animals stay and survive, some die, and some migrate. In the K–4 lesson, students learn through a read-aloud about 10 different animals that migrate, map their migration routes, and discuss reasons for migration. It is important for students to know that not all animals migrate, that there are other ways animals deal with changes in the environment. So in this lesson, students research how local animals respond to the changing seasons. Students in grades 5–6 study behavioral adaptations in more depth by addressing migration as a response to internal or environmental stimuli. In the lesson for grades 5–6, groups of students choose an animal from *Great Migrations*, research the migration patterns of that animal, share their findings, and discuss the internal cues and external cues that animals use to determine when it is time to migrate. Then, students join with another group to create a Venn diagram comparing and contrasting two migrating species. Both lessons end with a discussion of what is known about migration and what still remains a mystery.

Grades K–4: Going Home

Materials

- *Going Home* animal bookmarks
- World maps
- Colored pencils or markers
- Pictures of local animals (some that migrate and some that don't migrate)

Engage

Before class, print out several sets of bookmarks with pictures of the 10 animals featured in *Going Home: The Mystery of Animal Migration,* available on the Dawn Publications website (see Internet Resources). Give each group of three to four students a set of bookmarks. Show students the cover of *Going Home* and read the first page aloud. Tell them that as you read the rest of the book, you are not going to show them the pictures because you would like them to infer from the text which animal you are reading about. After reading the left-hand page for each animal, ask students to hold up the bookmark with the picture of the animal they think you are describing. Then, read the right-hand side of the page about that animal and show them the picture.

Explore

Provide each pair of students with a simple world map and several markers or colored pencils. Read the section in the back of *Going Home* titled "Find Their Route," which describes the migration route of each of the animals featured in the book. Have students find the route of one of the animals described in the reading and show the migration path that animal takes by drawing a double arrow in one color. Repeat for each of the animals using a different-color arrow, and then have them create a key that shows what each color of arrow represents. For younger students, you may want to project the map and do this together. When they have all the animals' migration paths on the maps, have a discussion about how the routes compare, which is the longest and which is the shortest.

Explain

Read the section in the back of *Going Home* titled "About the Migrating Animals," and ask the students to listen for the reasons why each animal migrates. Then, have groups of students sort the animal bookmarks by the reason for migrating (e.g., finding food, laying eggs, finding warmer temperatures). Ask, "Do any fit in more than one category?" Students will realize that some animals, like the manatees, migrate to be in warmer water and to find more food. Next, have them sort their animal bookmarks into groups based on how they travel (e.g., fly, swim, walk). Last, students can

The Mystery of Migration

create some of their own categories for sorting the animals.

Elaborate

Ask students whether any of the animals featured in *Going Home* live any part of their lives in your area. Research the animals in your area that migrate. Your state department of natural resources website might be a good place to start. It is important for students to understand that not all animals migrate—some hibernate, become dormant, or have adaptations that allow them to continue to live in the same place during the changing seasons. Print out pictures of some of your local animals and have students sort them into two categories: Migrates or Does Not Migrate. Map the migrations of some local animals and discuss why they migrate.

Evaluate

Read the section in the back of the book titled "The Mystery of Migration." As you read, ask students to listen for the reason why the author calls migration a mystery. Discuss what they have learned about migration and what questions they still have. Discuss the migration mysteries that scientists are still learning about such as How do animals know when it is time to migrate? and How do they find their way?

Grades 5–6: Great Migrations

Materials

- Animal migration poster scoring sheet (p. 183)
- Poster paper
- Sticky notes
- Copy of *Great Migrations* or laminated two-page spread (one per group)

Engage

Make a list on the board of the following animals: Mali elephants, red crabs, monarch butterflies, jellyfish, zebras, army ants, wildebeests, and sperm whales. Then ask students what the animals might have in common. Post photos of these animals where all can see, and allow students time to turn and talk to a partner about their ideas. Have students share their ideas with the whole class, then show them the cover of *Great Migrations*. Tell them that all of those animals migrate, or make a regular journey from one place to another. Then read aloud pages 10 and 11 about the *Great Migrations* television series, on which this book is based. Tell students that they are going to work in groups to become experts about the migration of one of the animals in the book. Read each two-page photo spread aloud, skipping the informational pages in between. You may want to leave out the name of each animal as you read and have students infer which animal is being described. After you read, have each student write his or her name and top three animal choices on a slip of paper. Use these to form research groups.

Explore

Provide each group with the animal migration scoring sheet and a copy of *Great Migrations* (or a laminated two-page spread of their animal). Tell them that they will be collecting a variety of information about their animals to share on a poster, but the big question they should be thinking about is, How do they know when to go? Students should use the book as well as the National Geographic Great Migrations website, which contains amazing videos and additional information about animal migrations (see Internet Resources).

Explain

Have students display their posters around the classroom or in the hallway. Set up a gallery walk in which they circulate around the room to view the posters. For each poster, have students write a comment and a question on a sticky note and place it on that poster. After the gallery walk, the poster groups can answer some of the questions they received. Tell students that one of the mysteries of animal migration is how the animals know when it is time to migrate. Create a T-chart labeled

"External Migration Cues" in the first column and "Internal Migration Cues" in the second column. With the class, make a list of external cues that tell animals when to migrate (e.g., amount of sunlight, changing seasons, and food or water availability). Then make a list of internal cues (e.g., fat reserves, hunger, thirst, or hormonal changes). Ask students whether the animal they researched uses internal cues or external cues or both. They will find that many animals use both.

Elaborate

Next, have each group of students combine with another group to create a large Venn diagram that compares and contrasts the animals each group has been researching. This will give each group a chance to teach some of what they have learned and compare their animal's migration pattern to other animals. Post these Venn diagrams in the room.

Evaluate

Assess student learning on their posters and Venn diagrams using the grading rubric on the following page. Last, discuss the idea that many aspects of migration remain a mystery to scientists. Ask students what they are still wondering about migration. Have each student jot down a question they have, then have each group choose the most compelling question at their table to discuss.

Reference

National Research Council (NRC). 1996. *National science education standards.* Washington, DC: National Academies Press.

Internet Resources

Educator Resources
    www.nationalgeographic.com/great-migrations-educator-resources

Going Home Animal Bookmarks
    http://dawnpub.com/activities/ GoingHomeBookmarks.pdf

The Mystery of Migration

Animal Migration Poster
4-3-2-1 Poster Grading Rubric

Names: _____

Your poster includes:

4 points: a detailed explanation of the internal and external cues that trigger your animal to migrate

          1   **2**   **3**   **4**

3 points: a picture of your animal and a labeled map showing its migration route, including the distance traveled in miles

          1   **2**   **3**

2 points: an explanation of the reasons for your animal's migration

          1   **2**

1 point: a description of some of the dangers your animal faces on its migration

          1

For extra credit, find a video of your animal's migration on the National Geographic website to share with the class *www.nationalgeographic.com/great-migrations-educator-resources.*

          1

Total Points _____/10

Comments: _____

Chapter 34

Whoo Eats What?

By Emily Morgan and Karen Ansberry

Owls are fascinating birds of prey. These nocturnal hunters are sometimes heard but rarely seen, making them even more intriguing and mysterious. These lessons feature two stories about owls. In both lessons, students dissect owl pellets to learn more about an owl's diet. The stories, along with the pellet dissection, provide a context for students to learn about energy transfer from one living thing to another.

Trade Books

Whoo Goes There?
By Jennifer Ericsson, illustrated by Bert Kitchen
Roaring Brook Press, 2009
ISBN 978-1-59643-371-7
Grades K–3

SYNOPSIS
With simple rhythmic, repetitive text, Jennifer Ericsson tells the story of one hungry owl waiting, listening, and watching for his dinner. Bert Kitchen's remarkable paintings capture the beauty and mystery of this nighttime hunter and some of the other nocturnal animals in its habitat.

White Owl, Barn Owl
By Nicola Davies, illustrated by Michael Foreman
Candlewick Press, 2009
ISBN 978-0-7636-4143-6
Grades 4–6

SYNOPSIS
This dual-purpose book tells the story of a girl and her grandfather waiting night after night to see a barn owl in the nest box they placed in a tree. Author Nicola Davies shares information about barn owls—such as their diet, adaptations, and how and why they make pellets.

Curricular Connections

These lessons provide literature links to owl pellet dissection, an activity done by students of many ages across the country. The stories featured here are meant to give students a context to what they are learning about—the flow of energy through living things.

The *National Science Education Standards* recommend that students in grades K–4 understand that all animals depend on plants; some animals eat plants, and some animals eat animals that eat plants (NRC 1996). Therefore, the focus of this lesson for K–2 is simply that an animal's food can always be traced back to plants. The Standards expand on this concept in grades 5–8 by adding the more abstract concept that sunlight entering ecosystems is transferred by plants into chemical energy through photosynthesis and that energy then passes from organism to organism through food webs. The lesson for grades 4–6 focuses on tracing the energy from the Sun through living things, using an owl food chain as the example.

Grades K–2: Whoo Goes There?

Materials

- Owl pellet dissection kit, available from science suppliers (see Internet Resource)
- Photos of owl prey found in pellets
- Picture cards for making food chains

Engage

Show students the cover of *Whoo Goes There?* Ask them what they think this book might be about and why there's an extra *o* in the title. Some students will recognize that "whoo" is the sound that some owls make. Next, read aloud the first page and ask students to turn and talk with a partner about what they think owls eat. As you read the rest of the book aloud, invite students to repeat the line "Whoo goes there?" each time it appears in the book. Mention that although the owl in the story didn't catch the cat, the skunk, or the porcupine, there are some kinds of owls that do eat those animals.

Explore

Tell students that owls swallow the animals they eat whole and then spit out the fur, feathers, and bones of those animals in a pellet. Tell students that they can find out what the owl that made a pellet ate by looking at the remains inside. Give each pair of students an owl pellet and have them separate the bones from the fur or feathers. See note at end of chapter regarding safety procedures. Help students use the skulls and a bone chart to identify the animals that the owl ate. (Bone charts are included with owl pellet orders from most vendors.) Next, make a list of the different animals that were found in the pellets and print a photograph of each one. Create a table with the photos in one column and the total number of each animal found in the class's pellets in the other column.

Depending on the age of your students, you could bring in parent volunteers or older students to help dissect the owl pellets and match the skulls to the bone chart. If volunteers are not available, or for children younger than grade 2, you could do the dissections as a demonstration—helping students access the bones and then making the match.

Explain

Next, reread the last 10 pages of the book, in which the owl pursues the mouse. Ask students what they think a mouse's meal might be. Have them turn and talk to a partner. Tell students that mice eat many different things, but the mouse in the book might eat seeds from the grassy field pictured on page 33. Remind them that seeds are parts of plants. Explain that the mouse can get food by eating the seeds and the owl can get food by eating the mouse. Create a simple food chain on the board. Draw an arrow from the seed to the mouse and from the mouse to the owl. Then draw the Sun above the diagram. Point out that this food chain begins with something

Whoo Eats What?

that came from a plant. Introduce the idea that all food chains include light from the Sun and begin with plants or plant parts. Plants do not eat other living things. Instead, they make their own food from air and water using light from the Sun.

Elaborate/Evaluate

Make five picture cards with an owl on each one. Then create picture cards with the other animals from the book. Make five more picture cards with an example of a plant or plant part that each animal might eat. Provide a picture of the entire plant if possible: nut from a nut tree (squirrel), clover (rabbit), seeds hanging on a plant (bird), berries from a bush (opossum), and water plant (fish). Together or in small groups, create simple food chains using the cards. Explain to students that these animals don't eat just one kind of food; they could eat several different kinds of food, and some could even eat other animals. Post all of the food chains in the classroom and ask students whether they notice what the pictures at the beginning of the food chains have in common. They should recognize that there is a plant or plant part at the beginning of all of the food chains.

Grades 4–6: Pass the Energy

Materials

- Owl pellet dissection kit, available from science suppliers (see Internet Resource)
- Hand lenses
- Scales
- Rulers
- My Mystery Object O-W-L chart (p. 189)

Engage

Do not tell students what owl pellets are prior to the lesson. Begin by saying that you have a mystery for them. Your friend has a farm in the country, and he keeps finding some strange things in his barn—can the class help him figure out their origin? Pass out the unwrapped owl pellets. Have each student write

a secret guess about what these objects are and why they made that guess. Collect the responses. By quickly reading through the student responses, you can determine what they already know about the objects without taking away from the excitement of discovery.

Explore

Give each student an O-W-L (Observations-Wonderings-Learnings) chart (p. 189). Have balances and tape measures available for measuring mass, length, and circumference. Have hand lenses available for examining the objects. Ask students to draw the objects and record their observations of the pellets in the O column. (See note at end of chapter regarding safety procedures.) Give them a few minutes to observe and measure the "mystery objects" without taking them apart. Encourage students to write quantitative as well as qualitative observations. Then ask students to write their wonderings about the mystery objects in the second column. Tell students that *White Owl, Barn Owl* might give them some clues about the mystery objects, and read it aloud. By merging clues from the mystery objects and the text and illustrations on pages 8–11, students should infer that the mystery objects are owl pellets. Tell students that owl pellets are the undigested remains of food eaten by owls that are regurgitated as a compact mass of hair, feathers, bones, teeth, claws, bird beaks, and insect remains. They can now add this learning to the *L* column of the O-W-L chart. At this point, encourage students to add more wonderings.

Explain

Have students separate the bones from the fur or feathers, and provide bone charts for them to use to determine what the owls ate. (Bone charts are included with owl pellet orders from most vendors.) After students have determined what animals their owls ate, ask them to research what *those* animals eat. Explain to students that when one animal eats another animal or plant, they both become part of a food chain. A food chain is the path that energy takes as one living thing eats another. Arrows are used to show the direction of the energy flow. On the board, create a simple food chain using an owl, one of the animals found in the pellet, what that animal ate, and so on until you can trace the energy back to a plant. Then draw the Sun above the diagram. Explain that plants don't eat other living things for energy; instead they use sunlight, carbon dioxide gas from the air, and water to make their own food for energy. Plants are called *producers* because they produce, or make, their own food to get energy for life processes. Animals are called *consumers* because they consume, or eat, other living things to get their energy.

Elaborate/Evaluate

Challenge groups of 3–4 students to come up with a model that tells the story of how some of the energy that originated from the Sun was eventually transferred to the owl that made the pellet they dissected. They should include the Sun, an animal the owl ate, and the other living things that may have been part of that particular food chain (including a plant). They can make a computer model, act it out, or create a piece of art. They should include the words *producer* and *consumer* where appropriate when they share their model.

Reference

National Research Council (NRC). 1996. *National science education standards*. Washington, DC: National Academies Press.

Internet Resource

Owl Resource Guide (including bone charts) www.carolina.com

Find out whether any students have fur or feather allergies. Although each owl pellet has been sanitized and is safe to dissect without using gloves, students must still wash hands with soap and water when finished.

CAUTION

Name: _____

My Mystery Object

O	W	L
What do you **OBSERVE** about the object?	What do you **WONDER** about the object?	What did you **LEARN** about the object?

Chapter 35

A Habitat Is a Home

By Christine Anne Royce

We all have our own habitats, and these activities allow students to spend time thinking about what other organisms need to survive, what types of habitats they live in, and how to set up a habitat for a classroom animal.

Trade Books

I See a Kookaburra!
Discovering Animal Habitats Around the World
By Steve Jenkins and Robin Page
Houghton Mifflin Books for Children, 2005
ISBN 978-0-618-50764-1
Grades K–3

SYNOPSIS
This colorful book employs the use of collages to present a hide-and-seek approach to finding animals that live in six different habitats. Following each collage, students are introduced to the animal inhabitants.

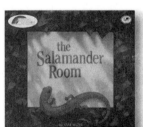

The Salamander Room
By Anne Mazer, illustrated by Steve Johnson and Lou Fancher
Dragonfly, 1994
ISBN 978-0-679-86187-4
Grades 2–5

SYNOPSIS
A young boy named Brian is asked a series of questions by his mother when he announces he wants to keep a salamander in his room. Questions such as "Where will he sleep?" prompt the young boy to come up with imaginative answers in an attempt to provide the proper habitat for his new salamander.

Curricular Connections

Each of us has our own "space"—the place we live that includes the things we need for survival such as food and shelter. Organisms adapt to living in their own habitats, and these modifications in structure or behavior happen over many lifetimes. However, many children may believe that animals are able to adapt to their environments in a short period of time.

At the K–4 level, the focus should be on understanding how organisms live in their environments. In the first activity, students make observations about animals they aren't familiar with. They begin to see that "organisms can survive only in environments in which their needs can be met" (NRC 1996, p. 129). Students then make the connection that this occurs in their local habitats, too.

The activity for older students allows them to take an in-depth look at a single ecosystem for a single organism through reading *The Salamander Room*. This helps develop an "understand[ing] of ecosystems and the interactions between organisms and environments" (NRC 1996, p. 156). Students then take their new knowledge and investigate how to create a habitat for a different animal. The activities provide a foundation for future development of these concepts.

Grades K–3: Hide and Seek

Purpose

To make observations about animals that live in different types of habitats

Materials

- Habitat Comparison Chart (p. 195)
- Chart paper
- Markers
- Reference books on local habitats
- Craft materials (such as glue sticks and colored paper)
- Magazines with pictures of animals

Procedure

1. Begin by writing the following words on the board and asking the students to describe their understanding of each word: *habitat, adapt, animal*.

2. Introduce the book *I See a Kookaburra!* to the class and explain that they are going to examine habitats from six different areas around the world (desert, tide pool, jungle, savanna, forest, and pond).

3. Give students copies of the Habitat Comparison Chart (p. 195). Pose the following questions to the students as they begin to fill in their initial ideas about each habitat: "Can you describe the [insert name of habitat]?" "Is it cold, wet, dry?" "What types of animals do you think will live in that habitat? Why?"

4. After students have had an opportunity to share their prior knowledge, begin by "reading" the book to them. Open to the first two-page spread, which shows a collage that says, "In the desert, I see ... While the teacher allows the students to make observations about the picture, students should name the different animals that are hiding in that habitat. Students might not necessarily have the exact name of the animal but may come up with something close. For example, in the desert, a Gila Monster is hiding on the right-hand side; the students might refer to it as a lizard rather than a Gila monster, but these types of answers should be accepted at this point.

5. Continue through the book, filling in student findings about the habitats and animals that live there on the comparison chart. Questions the teacher can pose as the students find an animal include "Can you describe what the animal is doing?" "What color is the animal, or what does the animal look like?" "Why do you think that helps the animal stay hidden in their habitat?"

A Habitat Is a Home

6. Once animals have been found, move onto the second two-page spread, which introduces the reader to each animal and provides the accurate name along with information about what they are doing (e.g., a long-nosed bat sipping nectar from a flower). Have the students compare their predictions and observations to what is presented, and ask students to think about what types of adaptations these animals have for living in that habitat. An example might be, "The bat has a long nose, which helps it get into the center of a flower to sip the nectar."

7. After the students have had an opportunity to explore habitats from around the world, introduce a local habitat to them—it may be a meadow, prairie, forest, beach, or whatever works for your location. If this is a new habitat from those discussed before, it should be added to the comparison chart. Take students outside to this area (if possible). Ask them to make observations about the local habitat: "What kind of animals live in this area?" "Where are their homes?" "What do they eat?" Students can then describe how the local habitat is similar to and different from the habitats presented in the book.

8. Engage students in research by asking them to use other books to look up information about the types of animals that would live in this habitat. Once they have found information on the animals, have them create their own collage (in drawings and/or cut-out pictures) that shows the animals in the habitat.

Grades 4–6: Building a Habitat

Purpose

To research the needs of an animal and then design an ecosystem for it

Materials

- Internet access
- Reference books (These resources will vary according to the animals selected to research.)

Procedure

1. Read *The Salamander Room* to the students once through without asking any questions. At the end of the story, ask students what happened in the story and to explain how the terms *habitat* and *ecosystem* are connected to the story.

2. After an initial discussion about habitats and ecosystems, reread the story to the students, allowing them to examine the pictures in more detail. While reading, ask the students, "Why is the room okay for the boy to live in but not the salamander?" "What parts of the room make it a 'boy's home'?" "What things do humans need to survive? Salamanders?"

3. As the story progresses, the boy's mother asks a series of questions about the salamander, such as "Where will he sleep?" "What will he eat?" "Where will he play?" Ask the students to respond to each of the answers provided by the boy. For example, the answer to "What will he eat?" is "insects." Students should explain that the food choice of salamanders is insects. This will help students connect the needs of the salamander to the requirements that an ecosystem must have to support that particular organism.

4. After students have explored what a salamander needs to live, they can investigate the needs of other animals and how to set up a habitat for those animals. (With your guidance, students could then set up a real classroom habitat for any of the following animals: salamanders, newts, turtles, fish, or lizards. See Internet Resources for information about the care of these animals. Follow your district guidelines and state regulations if you choose to have live animals in the classroom. Also, determine if

any of your students have allergies that would prevent you from housing a specific animal in your room.)

5. After deciding what type of animal habitat to research, students should begin to record the questions posed by the boy's mother in the story (e.g., "What will it eat?" and "Where will it sleep?") Finding answers to these questions will make students aware of the different needs for the particular animal's habitat and ecosystem.

6. Once the research has been finished, the students should present a detailed plan for a habitat for their selected animal including a description of how its needs are met.

Reference

National Research Council (NRC). 1996. National science education standards. Washington, DC: National Academies Press.

Internet Resources

National Geographic Kids: Animals and Pets
http://kids.nationalgeographic.com/Animals

NSTA Position Statement: Responsible Use of Live Animals and Dissection in the Science Classroom
www.nsta.org/pdfs/PositionStatement_LiveAnimalsAndDissection.pdf

Petco Animal Care Sheets
www.petco.com/CareSheets/petco_Nav_154.aspx?CoreCat-LN_PetCareInfo_AnimalCareSheets

A Habitat Is a Home

Habitat Comparison Chart

The habitat our class is going to examine is _____

_____

	DESERT	TIDE POOL	JUNGLE	SAVANNA	FOREST	POND
Description of Habitat						
Temperature						
Animals That Live There						
Why Do You Think the Animal Can Stay Hidden in the Habitat?						

Chapter 36

Exploring Your Environment

By Christine Anne Royce

Depending on your location, you may get to experience different seasons—thus meaning your environment changes. Students can begin to make initial observations about an area and then follow this area through the changes that occur over the course of several months, perhaps throughout the seasons. These activities focus on exploring the environment no matter where one resides—country, city, or suburb.

Trade Books

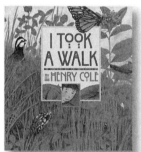

I Took a Walk
By Henry Cole
Greenwillow, 1998
ISBN 978-0-688-15115-7
Grades K–4

SYNOPSIS
A young boy takes a walk through the woods on a spring day. During his journey, he makes observations about the different animals, birds, and insects he sees. Pages in this book feature exquisite paintings that fold out to double-page spreads, allowing readers to make their own observations about these animals.

Secret Place
By Eve Bunting, illustrated by Ted Rand
Clarion Books, 1996
ISBN 978-0-395-64367-9
Grades K–4

SYNOPSIS
Secret Place is a beautifully illustrated book that tells the fictional story of a young boy who finds an area of outdoor tranquility in the heart of a big city. The book begins by describing the massive buildings, freeways, and smokestacks that are found near where he lives. However, through careful observation, the young boy finds what he describes as a concrete river, where a variety of different birds and animals come to nest and drink throughout the day. Other characters in the book have noticed this special place as well and explain, "Before the city grew there was wilderness." The young boy refers to this place of quiet as his "secret place."

Curricular Connections

These activities take students out into their environment—whether it's city streets, neighborhood parks, country fields, or carefully shaped and manicured suburbs. The children in both books use the science-process skills of observation and communication during their walks and when they describe what they find on their journey.

In addition to helping students develop these science-process skills, which are integral parts of any inquiry-based science program and need to be practiced, the activities help incorporate environmental appreciation into the curriculum. The topic of environmental awareness provides the opportunity for students to learn science content knowledge ranging from birds to flowers to habitats.

As the teacher reads either book to the students, there are opportunities for students to make additional observations and predictions based on the illustrations and prompting text. For example, students could make predictions about the animals that visit the river and observations about what they see in the pictures of the city, or they could compare the animals that visit a meadow to those that visit a wooded area. Providing students with these opportunities not only improves science-process skills but can also improve reading skills as well. After students have discussed the books, they can participate in one of the following investigations designed to help students move from observing the locations in the book to exploring their own secret place outdoors in their neighborhood.

Grades K–3: Rainbow Hunt

Purpose

Students use their senses to explore natural settings and create field journals.

Materials

- My Nature Notebook (p. 201)
- 2 in. × 2 in. pieces of different-color construction paper (one per student)
- Colored pencils

Procedure

1. Begin by asking students to describe the different objects they saw on their way to school. Some students may surprise you with the amount of detail they can recollect about something they saw, while others will have little description of the objects.

2. Pose the following question: "If we wanted to make observations about an area over time to explore how it changes, what are some ways that we could gather the information?" Students may offer suggestions such as "Take pictures of the area" or "Go back and look at it over time." Although both of these are valid suggestions, guide students to the idea that many scientists or naturalists take field notes and make observations.

3. Read *I Took a Walk* and ask students to make observations that answer the prompts posed by the author. Ask, "Do you think interesting things like that are found near us? Would you be interested in looking for different things on a walk like the young boy?"

4. Before venturing outdoors, introduce the idea that there are many objects that students can observe and that many of the objects come in many beautiful colors. (Teachers should check with their district policies on taking students outside or away from the school grounds during the school day before doing this activity.)

5. Explain that students will be making observations about objects in nature and drawing them in as much detail as possible on the My Nature Notebook student page (p. 201). Have each child choose one colored square of paper from

Exploring Your Environment

the box. This will be the color of objects that they are looking for on their walk. If a student selects a red piece of paper, he or she looks for red objects or objects with red in them, such as flowers or butterflies. This will take some patience and keen observation skills on the part of the student and enough time provided by the teacher for students to make observations and sketches. See student worksheet (p. 201).

6. It may take more than one trip outside for students to master their observation skills, but as students venture into their environment at different times, they are likely to see changes happen.

7. When children return to the classroom with their drawings, time should be provided for students to individually describe to the class what they observed. The drawing and describing of the objects help with the science-process skill of communication. This can be taken a step further by having all students with red squares get together and determine all of the red objects observed.

8. On subsequent trips outdoors, change the focus for the students from colors to something else—such as size of objects. Or students could lie on their backs and make observations about what is above them.

9. Now that students have drawings from their outdoor adventures, there are many additional activities that can be developed using the other science-process skills, such as having each student choose one favorite thing they observed and draw it on an index card, and then having the entire class create a bar graph on the wall.

Grades 4–6: Finding Your Own Special Place

Purpose

Students will examine a habitat over time to describe how it changes.

Materials

- My Nature Notebook (p. 201)
- Colored pencils

Procedure

1. Older students can take the previous activity one step further and begin to make observations about one place over time to determine how it changes.

2. Locate an area close to the school that students will be able to visit several times during the school year. If this is impossible to do as a class, students could be asked to locate a place near their homes and complete their observations outside of school time.

3. After reading *Secret Place*, introduce students to the idea that they will be making observations about their own special place over time as well as examining a different habitat. Questions that help the students prepare include the following: "What are the components we would want in our field notes to help us know when we wrote them?" (dates, times, location). "When making observations, how can we record the information we gathered?" (illustrations, descriptive phrases, sketches, notes). "What are some things you would expect to see in the habitat that we will visit?"

4. Teachers should provide ample time for students to investigate their special place and make field notes and drawings about what they see, hear, smell, and can touch. Because students will draw sketches of what they see to make comparisons, they should be as detailed as possible. Questions such as, "Has anyone looked up while standing in your special place?" and "What do you notice?" help students make observations from different points of view.

5. Upon returning to the classroom, students should be once again provided with an oppor-

tunity to share their observations with their classmates through illustrations, notes, verbal descriptions, and so on. Their observations and sketches of their special place will change over time as the place will change, but the students will also become more detailed about what they observe. Teachers could take the opportunity to set aside a corner of the classroom to display the "special places" the students found.

6. Using the data that the students have collected, ask them to define what type of habitat they found: "Is it aquatic or terrestrial?" "What types of organisms did they observe living in the habitat?" "How do they interact?" Students can use research materials along with their field notes to develop an informational sheet for their habitat.

7. This activity does not stop at the science classroom door but rather can be expanded to other subject areas. The information obtained in the field notes and sketches made by the students can be used in language arts lessons through writing samples, poems, and even letters requesting further information about the park or other area from local sources.

8. A culminating activity could be the creation of an illustrated book similar to *Secret Place* that uses the student's notes and sketches.

The awe of the outdoors brings with it many opportunities to expand your classroom walls and your science lessons into other subject areas while your students learn to appreciate the environment.

Reference

National Research Council (NRC). 1996. *National science education standards.* Washington, DC: National Academies Press.

Exploring Your Environment

My Learnings

Sketch the different objects you see or the place you visit:

My Nature Notebook

Things you "see" or observe in your environment:

Chapter 37

You Are What You Eat!

By Christine Anne Royce

Kids today have a dizzying array of food choices, but making healthy choices is a challenge. Making wise choices is essential in maintaining a healthy lifestyle. The topic of healthy foods fits into science nicely when students start to consider the value of what they eat and experiment to determine the makeup of some of their favorite foods.

Trade Books

Gregory, the Terrible Eater

By Mitchell Sharmat, illustrated by Jose Aruego
and Ariana Dewey
Scholastic Trade, 2009
ISBN 978-0-545-12931-2
Grades K–3

SYNOPSIS
Gregory the Goat is a "terrible eater" from a goat's perspective. He refuses to eat items such as tin cans and shoes, which are staples in a fictional goat's diet. Rather, he wants to eat healthy things such as eggs, fruits, and vegetables. The illustrations help students see that they have a choice in selecting what they eat—even if it isn't a popular choice.

Science News for Kids: Food and Nutrition

Edited by Tara Koellhoffer
Chelsea Clubhouse, 2006
ISBN 978-0-7910-9121-0
Grades 3–5

SYNOPSIS
This book is part of a series that highlights news articles for students and is a great springboard to discussing a common topic—what we eat and, more important, how what we eat affects our lives. The book uses a variety of strategies—articles, sidebars, illustrations, and pictures—to provide information on the topic of how to select healthy foods for a healthy lifestyle.

Curricular Connections

In many elementary schools, health and science are taught together. Making healthy food choices also falls under the Science in Personal and Social Perspectives standard, which states, "Nutrition is essential to health. ... Recommendations for good nutrition include eating a variety of foods, eating less sugar and eating less fat" (NRC 1996, p. 140). These activities inform students about healthy eating habits, but it is important to treat this topic with sensitivity and to not pass judgment about what students may be eating at home. This activity is meant to provide students with the opportunity to explore what healthy foods are in order to make informed decisions. The best lesson is to eat a varied diet—there are many choices and many kinds of food can be considered healthy.

Grades K–3: The Food Pyramid

Purpose

Students will learn about the different types of food found on the food pyramid and then identify foods that would fall into each of those categories.

Materials

- Food Pyramid handout (see Internet Resources)
- Newspaper inserts from grocery stores
- Other pictures of food
- Glue sticks

Procedure

1. Pose the question, "Do different animals eat different things?" Most students will respond yes. Next, ask, "Do you think certain foods are good for different animals? For example, do you think chocolate is good for fish?" Most will say that fish eat fish food or maybe other fish but not chocolate. After introducing the ideas that different animals (including humans) eat different things and that

some things are good or healthy for the individual, read *Gregory, the Terrible Eater* to the class.

2. Stop at page 5 and ask the students whether they would like to eat "a tin can or a rug or an old shoe" like Gregory the Goat's parents suggest. "Why wouldn't you like to eat these things? Do you think they are healthy for you?" Continue reading the story as Gregory begins to eat junk (literally) and then doesn't feel well. Pose the questions, "Are there foods that make you feel better after eating them? Are there foods that make your stomach hurt?" Has someone ever said, "If you eat too much of that food you will get a stomachache?"

3. Have students brainstorm a list of what they eat on a regular basis, naming specific items such as cereal, peanut butter, and potatoes. While they list these items, post the following questions: "Which items that we have listed are healthy for boys and girls? Why do you think that?" Have students explain their reasoning, such as, "I eat yogurt, which is good for us because it is a milk product and we need milk for strong bones."

4. Present the food pyramid to the class (see Internet Resources). Ask students to classify the foods that Gregory the Goat ate into the different categories. "Where do eggs fall? Do pants and shirts fall anywhere on this list?" Students tend to be amused about the idea of eating junk but realize that not everything the goat eats falls into the food pyramid categories.

5. Once the students have tackled Gregory's diet, have them classify their own food lists as a class. Not all of the foods they name will fit easily into a category—these tend to be prepackaged foods. Some students will start to see that many of the foods they eat are not healthy choices.

6. Provide students with a blank copy of the food pyramid and various food pictures. Ask them to sort the food pictures into each of the categories. This will allow them to begin to see the diversity

You Are What You Eat!

of different categories. For example, orange juice falls into the fruit group, peanut butter is a meat and bean group item (since it is a protein), and pasta falls into the grain category. (Obviously it is healthier to eat whole grain pastas than white flour pasta, but that is beyond the scope of this activity.) They can then paste the pictures onto that area of the food pyramid and color it. The teacher can extend this lesson by creating a class graph of the students' favorite foods in each of the categories.

While *Gregory, the Terrible Eater* is a humorous take on goat diets, it is important to make sure to clarify information before students walk away from the lesson with potential misconceptions. Goats are extremely adaptable in their dietary habits, but however varied this might be, it does not include tin cans. Through this activity, students will begin to see that like goats, they have a variety of healthy foods that they can choose from to meet their daily requirements.

Grades 4–6: Testing Foods

Purpose

To identify, through testing, which foods contain complex carbohydrates, sugars, fats, or proteins

Materials

- Paper towels
- White paper
- Iodine solution (and MSDS)
- Benedict's solution (and MSDS)
- Matches
- Candle
- Dissecting needle
- Food samples (per group of two to three students): 1 tsp cooked hamburger
- Cracker
- ½ in. slice of a potato
- Apple or pear
- Potato chip
- 1 in. × 1 in. square of cheese

Note: Students must wear nonlatex gloves, aprons, and indirectly vented chemical-splash goggles for the following activities. Nothing in this experiment should be put in or near the mouth—food in science class is for investigation only. The solutions and dissecting needles can be obtained from your district's high school biology class. During the demonstration involving fire, you must have an A-B-C type fire extinguisher in the room.

Procedure

1. Have students select one of the short articles from *Food and Nutrition* and present brief explanations of the main topic of the article to the class. Throughout the articles, there is a common theme: selecting healthy foods. As students provide overviews of the articles, the idea that students should eat a variety of foods including grains, proteins, carbohydrates, and oils—all in moderation—becomes evident.

2. Have the students put each of the foods on a paper towel. Students may need more than one sample of each food. Before starting, students individually or as a class should make predictions about the category of each food based on their readings and presentations from the book.

3. Before students begin, share information from the MSDS, review proper safety, and model the procedure. To test for starch-type (complex) carbohydrates, students should add one drop of iodine to each food sample. If there is a color change to black, a starch is present in the food. If the drop remains a reddish brown (the original color of the iodine), no starch is present. To test for sugar-type (simple) carbohydrates, students should add a single drop of Benedict's

solution to the sample (in a different spot) and once again watch for a color change. If there is a dark blue-green color change, sugars are present in the food. If the color is light blue-green, no sugar is present.

4. Using new samples of food, students should now test for fats by placing the food samples on a white sheet of paper. Using a sheet of paper on top of the samples, have them smash or mush the food samples onto the paper. Remove all of the food samples and inspect the paper for a translucent mark (the same type of mark that would be left from greasy fingers on a paper). If a translucent mark is evident, there was fat in the food; however, if the paper is simply wet, there is little or no fat present.

5. To test for proteins, the teacher should conduct a demonstration. Collect a small amount of each food item, one at a time, on the dissecting needle, and place it over a candle flame. If the food sample begins to burn, remove it from the flame, and blow it out. Using a wafting motion with your hand, ask students to determine whether there is a rotten egg smell (sulfur) present. If so, the sample contains proteins that are being broken down by the flame. If no odor is present, proteins are not present in the food. Never place smoking materials directly under your or your students' noses.

6. Upon completion of activities, students need to wash off tabletops and hands with soap and water. Be sure to properly dispose of the tested food items.

7. Refer back to the predictions made by the students and see how accurate they were. Were there any surprises?

Understanding what a healthy diet comprises is an important aspect of daily life. These activities will help students connect the science of food makeup to the abundant variety of nutritional choices available to them on a daily basis.

Reference

National Research Council (NRC). 1996. *National science education standards*. Washington, DC: National Academies Press.

Internet Resources

Food Pyramid handout
http://teamnutrition.usda.gov/resources/mpk_coloring.pdf

USDA Food Pyramid Poster
http://teamnutrition.usda.gov/Resources/mpk_poster2.pdf

USDA My Pyramid
http://mypyramid.gov

Chapter 38

Moving My Body

By Christine Anne Royce

The human body is amazing! The trade books highlighted here help students understand their growing bones, muscles, and joints while engaging in some investigations to help them become aware of their bodies and how exercise helps their bodies stay strong.

Trade Books

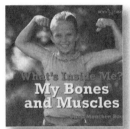

What's Inside Me? My Bones and Muscles
By Dana Meachen Rau
Benchmark, 2008
ISBN 978-0-7614-3351-4
Grades preK–3

SYNOPSIS
Young students are introduced to their bones and muscles through carefully worded text. This book is part of a series that helps answer the question, "What's Inside Me?" and is written at the level of the fluent reader. This book examines the skeletal system and the importance that muscles play in helping us move. Through the use of illustrations, italicized vocabulary, and pictures, students are provided with the basics about these two body systems.

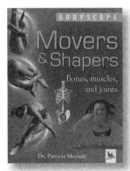

Movers and Shapers
By Patricia Macnair
Kingfisher, 2004
ISBN 978-0-7534-5791-7
Grades 4–6

SYNOPSIS
Written for the intermediate level student, this book connects understanding the anatomy of the human body with the importance of exercise and well-being. Through the use of illustrations, clear text, and a pullout poster, students are introduced to the roles that bones, muscles, and joints play in helping humans move and function. This book also includes suggested websites, a glossary, and an index.

Curricular Connections

Children are in constant motion. Teachers can capitalize on this movement and energy to engage students in wondering, "How do we move? What does motion or exercise do to the other parts of the body?"

Younger students can think about how their daily movements would be different without the use of their joints. Simple, daily tasks of picking up objects, walking, and sitting would be different if our bodies didn't have the flexibility and mobility available through the use of joints. By investigating how their lives would be different, students can begin to develop a concrete understanding about the parts of their body they cannot see.

Older students can be introduced to the idea that strong muscles are needed for a healthy life and that exercise can contribute to changes in our breathing rate and heart rate. These activities will have students recording data and drawing conclusions about the effect that exercise has on pulse rate and respiration.

Regardless of which activity students engage in, *What's Inside Me?* and *Movers and Shapers* will provide students with nonfiction resources they can refer to during an elementary-level anatomy unit.

Grades K–3: Trying Tasks

Purpose

Students will engage in activities to help them recognize how different joints in the human body are helpful in accomplishing daily tasks.

Materials

- Poker chips (15 per group)
- Pennies (15 per group)
- Box
- Masking tape
- Sponge ball (about 20 cm in diameter)
- Basketball
- Plastic laundry basket
- Ace bandage

Procedure

1. Begin by reading *What's Inside Me? My Bones and Muscles* to the class. Ask students to perform the actions suggested on page 4—stand still, bend over, wave your arms, and kick your legs—while covering the text on page 5. Ask, "What parts of your body help you do each of these things?"

2. As you continue to read the book, ask students to point to the location of the different bones mentioned. Explain how bones are inside our bodies and covered by muscles and skin. Record the italicized vocabulary words on chart paper or the board.

3. After completing the book, ask which word refers to the point where two bones meet (joint). "Where can we find different joints in our bodies?" While they are thinking about joints, ask, "What type of actions do joints allow us to perform every day?" (Bend knees, move shoulders, pick up objects with fingers and thumb).

4. Have students investigate how the different joints move—can students tell that different joints move in different ways? Compare the thumb to the elbow or knee. Develop a taxonomy of joints. The elbow is called a hinge joint. What else has hinges? How does it move (like a door or gate)?

5. Next, demonstrate how to immobilize the thumb and elbow joints, making sure to point out safety concerns—do not overextend a joint; when immobilizing the joint, do not wrap it too tightly; students should spot each other if necessary.

6. *Penny Pinch* (using the thumb joint): Ask students to think about how much they use their thumb in accomplishing daily tasks. Students should tape their right or left thumbs to the palm of their hand so that the thumb lies across their palm. Once the

Moving My Body

joint is immobilized, ask students to remove 15 poker chips from a box and neatly stack them in three groups of five on the table. Students should record their observations about this activity, describing if they found this task easy or difficult and why. Have them repeat this activity using pennies. Ask, "Does the size of the object make a difference?"

7. *Elbow Grease* (using the elbow joint): Either the teacher or an adult volunteer should take an Ace bandage wrap (the type you would use when you sprain an ankle) and wrap a student's arm so that they cannot bend their elbow. Do not wrap the Ace bandage too tightly as it can cut off their circulation to the arm. Using only this arm, each student should pick up a sponge ball and toss it into a basket that is placed 2 m away. Is it easy to pick up the ball? To toss the ball? Repeat this activity with a basketball (students may need to use both hands to pick up the basketball). Is it more difficult?

8. While students participate in each activity, ask the other students in the group to make observations: How is the person moving? How would they be moving if they could use their joint? How are they adjusting their movement since a joint is immobilized?

9. After all students have had the opportunity to participate, ask them to think of other joints such as finger joints or knees, and how not being able to bend these joints would affect their movement.

10. Many students will say that it was hard to do these things, or that they couldn't do these tasks. Other students will be able to describe why their elbow or thumb would help by stating that it is easy to bend their arm when they throw the ball or that their thumbs help them hold the poker chips. Then ask students to think about how scientists might study joints when they can't see inside. What kinds of tools can they

use? This discussion is a perfect opportunity to help students begin to understand more about what scientists do and the nature of science.

Grades 4–6: Pump Up Your Body!

Purpose

Students will investigate how exercise affects their breathing and heart rates.

Materials

- Stopwatch
- Pencil
- Large area such as a gymnasium
- Data Collection Chart (p. 211)

Procedure

1. Older students may begin with the "Trying Tasks" activities after reading pages 10–15 of *Movers and Shapers*. This could be an interesting way to explore how different parts of the body, such as joints, have different functions. Pose the following questions to students: "Why is it important to exercise? How does exercise help our body function?" Possible student predictions about how exercise affects your body may include increases heartbeat, breathing goes up, muscles get sore, and so on.

2. After the brainstorming session, read pages 34–35 to the class (about the need to be active and how exercise affects your body). Ask, "Do you think all types of exercise have the same effect? Why or why not?"

3. Explain to students that they will participate in an activity to see how exercise affects their heart rates. Explain that the heart is a muscle that pumps blood throughout the body. Have students select three different exercises to do in the school gymnasium. (It may even be a fun idea to enlist the assistance of your physical

education teacher to help with this activity.) The exercises should involve varying levels of exertion (e.g., sitting, walking, and jumping jacks).

4. Next, students should predict which exercise will increase their heartbeats the most and which one will increase their heartbeats the least. Record the predictions on the board.

5. Have the students practice taking their pulses on the inside of their wrists. Students should use their index and middle fingers, not their thumbs, to feel for a pulse. The students should practice recording their pulses for one minute. If students are not able to locate their pulses on their wrists, have them try to find it on their necks, slightly to the side.

6. Now that the students have selected their exercises and can take their pulses, have them gather data on the Data Collection Chart. Each exercise chosen should be done for at least three minutes. Students should allow their heart rates to return to normal before doing the next exercise. Students should be monitored during exercises and teachers should check with the appropriate office about health-related issues involving exercise.

7. Have students analyze the data by asking them, "Did all activities affect your hearts in the same way? Which activity increased your heart rates the most? Did it take the same amount of time for your heart rates to return to normal after each exercise?" Students may choose to graph their pulse rates against time on a large graph to show the difference between individuals. It is recommended that you exercise continuously for 30 minutes a day. Ask students which exercise they would choose and why.

After students have had an opportunity to rest, return to *Movers and Shapers* to identify the different body systems used during their exercises. Ask, "What was the purpose of the skeleton during exercise?" (to provide structure and a framework for the body). The activities successfully engage students in understanding the body!

Reference

National Research Council (NRC). 1996. *National science education standards.* Washington, DC: National Academies Press.

Moving My Body

Name: _____

Data Collection Chart

EXERCISE	PULSE BEFORE EXERCISE	PULSE IMMEDIATELY AFTER EXERCISE	PULSE 1 MINUTE AFTER EXERCISE	PULSE 3 MINUTES AFTER EXERCISE	PULSE 5 MINUTES AFTER EXERCISE

Chapter 39

Science From the Heart

By Karen Ansberry and Emily Morgan

*I*t's a fact: Kids today are less fit than they were only a generation ago. Many are showing early signs of cardiovascular risk factors such as physical inactivity, excess weight, and higher blood cholesterol. Now more than ever, it is important to teach children how to keep their hearts healthy. The following books and activities engage students in simple heart investigations and help them learn that regular physical activity can promote heart health.

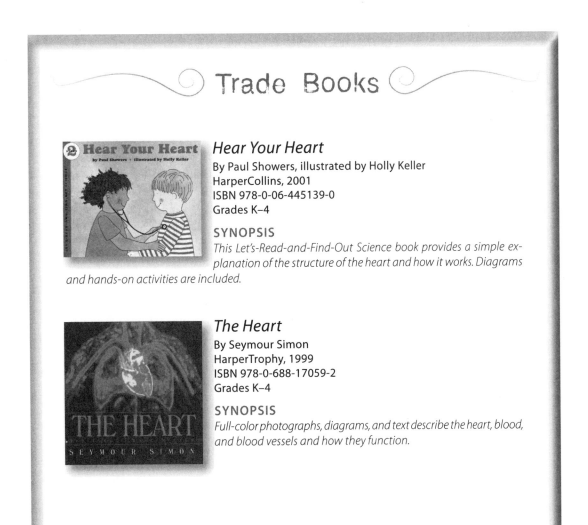

Trade Books

Hear Your Heart
By Paul Showers, illustrated by Holly Keller
HarperCollins, 2001
ISBN 978-0-06-445139-0
Grades K–4

SYNOPSIS
This Let's-Read-and-Find-Out Science book provides a simple explanation of the structure of the heart and how it works. Diagrams and hands-on activities are included.

The Heart
By Seymour Simon
HarperTrophy, 1999
ISBN 978-0-688-17059-2
Grades K–4

SYNOPSIS
Full-color photographs, diagrams, and text describe the heart, blood, and blood vessels and how they function.

Curricular Connections

The activities described here combine learning about the heart and circulatory system with lessons about personal health. In the K–3 lesson, students listen to each other's hearts, participate in a simple activity to find out how exercise affects heart rate, read about the heart in Paul Showers's *Hear Your Heart,* and make a graph of favorite heart healthy activities. In the 4–6 lesson, Seymour Simon's *The Heart* provides information to help students design their own heart-rate experiments.

Learning how to keep the heart healthy while learning how it works gives real-life context to the material. The heart contracts and relaxes automatically as it pumps blood to all parts of the body through an intricate system of blood vessels. A healthy heart makes a *lub-dub* sound with each beat. This sound comes from the valves shutting inside the heart. Blood leaves the left side of your heart and travels through blood vessels called *arteries*, which gradually divide into *capillaries*. Inside capillaries in the lungs, oxygen/carbon dioxide exchange occurs, and in the intestines, nutrient/waste exchange occurs. The blood then travels in veins back to the right side of your heart, and the whole process begins again.

You can feel each time the heart squeezes a jet of blood into the arteries by finding your *pulse.* Two good places to find it are on the side of your neck just below the chin (the carotid artery pulse) and on the inside of your wrist just below the thumb (the radial artery pulse). When you are resting, you will probably feel between 70 and 100 beats per minute. A child's resting heart rate is faster than an adult's.

Because the heart is a muscle, exercising it helps keep it healthy and strong. The American Heart Association recommends you do some sort of cardiovascular exercise for 30–60 minutes most days of the week. *Cardiovascular,* or *aerobic,* exercise is moderate exercise done for a long period of time that gets your heart rate up, such as running.

Grades K–3: Hearing Hearts

Materials

- Cardboard paper towel or toilet paper tubes (one per pair)
- Busy Body Cutouts (p. 217)
- Your Heart Anticipation Guide (p. 218)

Engage

Begin by asking students, "What is the most important part of your body?" "Why do you think so?" Responses will vary, but tell students that they will be learning about an organ they couldn't live without: the heart. Using a projected image of the Your Heart Anticipation Guide as a preassessment tool, read the statements about the heart (e.g., true or false: Your heart beats slower when you exercise) and have the class discuss whether they think each statement is true or false. Mark answers in the "Before" column of the anticipation guide. At the end of the lesson, students will complete the "After" column.

Explore/Explain

Have students look at their closed fists. Tell students that the human heart is about the size of a fist. Have them place their fists against their chests and explain that most of the heart is located a little left of center in the ribcage. Give each pair of students a cardboard paper-towel or toilet-paper tube "stethoscope." Have them take turns listening to each other's heartbeats by putting one end of the tube on the left side of their partner's chest and placing their ear to the other end. Make sure students are silent during this activity. After all students have had a chance to listen to their partner's heart, ask, "What does your partner's heart sound like?" "Do you think the sound will change if we exercise before listening?"

Have students do some sort of cardiovascular exercise, such as marching to music or running in

Science From the Heart

place, for one minute. Then have them repeat the cardboard stethoscope activity. Ask, "How does your partner's heart sound different after exercising?" Students will comment they hear heartbeats that are quicker or stronger.

Next, show students the book *Hear Your Heart,* and explain that it is a nonfiction book that can help them find out more about the heart. Read the book aloud, skipping pages 12–21 to focus on the portions of the book about the heartbeat and pulse. Then ask,

- "Is the heart a muscle?" (yes).
- "How do you keep your muscles strong?" (exercise them).
- "What can you do to keep your heart strong?" (exercise it by running, jumping, playing sports, and so on).
- "How often should you do these activities every week?" (most days for 30–60 minutes).

Elaborate

Make a large "Busy Body" graph by labeling the *x*-axis "Favorite Physical Activities" and the *y*-axis "Number of Students." Brainstorm a list of five or six cardiovascular activities that students enjoy and write them along the *x*-axis. Pass out "Busy Body" cutouts (p. 217) and have students write their names and their favorite activities from the choices listed. Next, have students tape their cutouts on the graph to create a pictograph of favorite activities and then analyze the graph together. Tell students that they can keep their hearts strong by keeping their bodies busy with fast-paced physical activity for 30–60 minutes most days of the week.

Evaluate

Revisit the anticipation guide. Have the class consider again whether each statement is true or false, and mark the answers in the "After" column (p. 218). Discuss evidence for each statement from the activities and the reading, such as, "Our hearts

are muscles so they need exercise like our other muscles."

Grades 4–6: What Gets the Heart Pumping?

Materials

- Your Heart Anticipation Guide (p. 219)
- Poster paper
- Markers
- Healthy Heart Investigation poster rubric (p. 221)

Engage

Have students quickly open and close their fists over and over again until their hands get tired. Tell them that their fist-size heart squeezes this way every second of every day.

Explore

Next, demonstrate how to measure your heart rate at your wrist or neck. Students may have to try a few different spots until they feel a soft beating. This beating is called a *pulse,* and it is caused by the heart squeezing blood through the body. Tell students that their pulses can tell them how fast their hearts are beating if they count the number of beats for one minute (or multiply how many times their hearts beat in 15 seconds by four). Students can graph their heart rates, compare the heart rates of different students, and find the average heart rate of their class.

Explain

Ask students to share with a partner something they know or have heard about the heart, such as, "The heart has four chambers in it" or "People can get artificial hearts." Pass out to each student a copy of Your Heart Anticipation Guide (p. 219) and have them fill in the "Before Reading" column. Then read aloud from the book *The Heart.* Read only pages 4–9 ("Make a fist ...") and page 22 ("The heart pushes ...") as this book is somewhat lengthy to read aloud cover-to-cover. Ask students to listen to find out if each statement on the anticipation guide is true or false. After reading, have students complete the "After Reading" columns, marking whether they agree or disagree with each statement, and provide evidence for their answers.

Elaborate

The American Heart Association recommends that we do two and a half hours of moderate cardiovascular exercise (or 75 minutes of vigorous exercise) each week. Have teams of students design investigations to determine ways they could get better cardiovascular exercise during their free time. For example, "In which activity is your heart beating the fastest: playing a hand-held video game, jogging, or jumping rope?" To help them understand the concept of a fair test, approve students' procedures before they begin carrying out the experiments. Ask, "How will you keep the experiment fair?" (doing each activity for the same amount of time, taking their pulses in the same way each time, and so on) and "How will you record and organize your data?" (writing it in a table).

Evaluate

Have students create a "Healthy Heart" poster. Give them the Healthy Heart Investigation poster rubric (p. 221), which explains the following criteria:

* The question the team was investigating.
* The procedure they used for the experiment.
* A data table or graph to show results.
* The team's conclusion and evidence to support it.
* A list of ways to keep the heart healthy.

Have students share their posters at a Healthy Heart poster session.

Reference

National Research Council (NRC). 1996. *National science education standards.* Washington, DC: National Academy Press.

Internet Resource

American Heart Association
www.americanheart.org

Science From the Heart

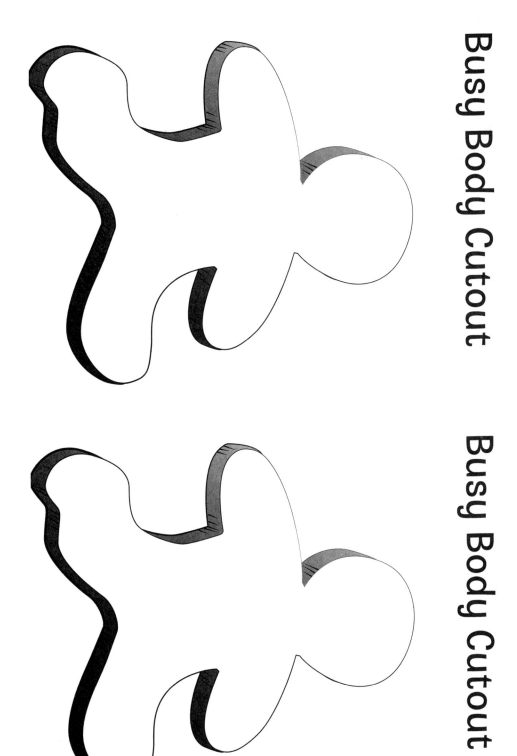

Busy Body Cutout

Busy Body Cutout

Name: _____

Your Heart

Anticipation Guide
(Grades K–3)

Before **True or False**		*After* **True or False**
_____	1. Your heart is closest to the right side of your chest.	_____
_____	2. Your heart is about as big as your fist.	_____
_____	3. Your heart is a strong muscle.	_____
_____	4. Your heart beats slower when you exercise.	_____
_____	5. A big animal's heart beats more slowly than a little animal's heart.	_____
_____	6. You should exercise for 30 to 60 minutes most days of the week.	_____

Science From the Heart

Name: _____

Your Heart
Anticipation Guide
(Grades 4–6)

Directions: Before reading the book *The Heart*, mark whether you agree or disagree with each statement by placing a check mark in one of the "Before Reading" columns. After reading, mark whether you agree or disagree with each statement by placing a check mark in one of the "After Reading" columns. Then use evidence from the reading to support your choices.

Before Reading   **After Reading**

AGREE	DISAGREE	STATEMENT	AGREE	DISAGREE
		1. Your heart beats every second of every day.		
		2. All animals have hearts.		
		3. Your blood brings food and oxygen to each cell and carries away carbon dioxide.		
		4. Your heart weighs about 10 pounds.		
		5. Each of the heart's four chambers has a one-way valve to keep blood from flowing backward.		
		6. Children's pulses are slower than adults' pulses.		

Evidence from the reading in your own words:

1. _____

2. _____

3. _____

4. _____

5. _____

Name: _____

Your Heart

Anticipation Guide
(Grades 4–6)

Directions: Before reading the book *The Heart*, mark whether you agree or disagree with each statement by placing a check mark in one of the "Before Reading" columns. After reading, mark whether you agree or disagree with each statement by placing a check mark in one of the "After Reading" columns. Then use evidence from the reading to support your choices.

Before Reading   **After Reading**

AGREE	DISAGREE	STATEMENT	AGREE	DISAGREE
		1. Your heart beats every second of every day.	✓	
		2. All animals have hearts.		
		3. Your blood brings food and oxygen to each cell and carries away carbon dioxide.	✓	✓
		4. Your heart weighs about 10 pounds.		✓
		5. Each of the heart's four chambers has a one-way valve to keep blood from flowing backward.	✓	
		6. Children's pulses are slower than adults' pulses.		✓

Evidence from the reading in your own words (sample responses):
1. Your heart is always beating.
2. Some tiny animals in oceans and ponds do not have hearts.
3. Your blood brings food and oxygen to your cells and takes waste away.
4. Your heart only weighs about 10 ounces.
5. The heart has four chambers. Each one has a valve that closes so blood can't flow backward.
6. Children's pulses are faster than adults' pulses.

Science From the Heart

Healthy Heart Investigation POSTER

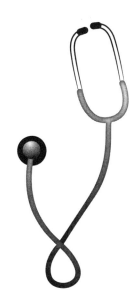

Make a poster with your team displaying the findings of your healthy heart investigation. Your poster should include the following elements:

- The question your team was investigating.

- The procedure you used for the experiment.

- A data table or graph of your results.

- Your team's conclusion and evidence to support it.

- A list of ways to keep your heart healthy.

- For extra credit: An advertisement, commercial, or jingle that teaches others about keeping their hearts healthy.

Decide who is going to explain each part of the poster. Everyone on your team should have a turn!

Posters will be shared on (date) _____.

Chapter 40

Disease Detectives

By Christine Anne Royce

"You've got cooties!" is a traditional taunt between children on the playground. Although children's avoidance of "boy" or "girl" germs is not scientifically based, students have an innate understanding that transmission of germs can happen through touching. Although these activities won't touch on "cooties," they will provide students with an opportunity to examine what germs are and how they are spread.

Trade Books

Germs Are Not for Sharing
By Elizabeth Verdick, illustrated by Marieka Heinlen
Free Spirit Publishing, 2006
ISBN 978-1-57542-196-4
Grades K–2

SYNOPSIS
The author provides a developmentally appropriate overview of how to prevent the spread of germs by explaining where germs can be found, how they can be transmitted to others, and different strategies for eliminating germs. The book glosses over the issue of germs making you sick, so consider how much this point should be emphasized.

What Are Germs?
By Alan Silverstein, Virginia Silverstein, and Laura Silverstein Nunn
Children's Press, 2003
ISBN 978-0-531-16640-6
Grades 3–8

SYNOPSIS
A wonderful overview of germs is presented throughout this text in clear and detailed narrative. Through the use of actual photographs, illustrations, and cartoons, the authors present information on what germs are, how your body fights germs, different types of germs that students may be familiar with, and how doctors help treat diseases caused by germs.

Curricular Connections

Almost all students have had either a cold or the flu or some type of childhood disease and most can describe what the illness "did" to them and how they felt (e.g., fever, sneezing). However, understanding "how" they got sick is another concept entirely as this includes a basic concept of the manner in which germs, and thus diseases, are spread from one human to another. Furthermore, children typically use *germs* to describe all microbes, be they *virus, fungi,* or *bacteria.* Even when they do use the correct vocabulary, they do not understand the distinctions (NRC 1996, p. 139). Consider for a moment how often we say "cold" or "flu," rather than "flu virus," when talking about the annual season of colds and the flu. A second idea that is developed when we discuss diseases is the distinction between contagious and noncontagious diseases. Although students are more familiar with contagious ones, as that is what we try to prevent, it is important for them to also understand that some diseases are genetic or contracted through environmental conditions. The *National Science Education Standards* state that "teachers can expect students to exhibit little understanding of ideas, such as different origins of disease, resistance to infection, and prevention and cure of disease" at the elementary level (NRC 1996, p. 139) Hopefully through participation in the activities described here, students will begin to develop basic conceptual understanding in these areas. In the first activity, students investigate that different types of microbes are floating in the air and what happens if they are allowed to grow on a substance. In the second activity, students take on the role of detective in trying to determine who was the "infected" patient during a simulation activity. The second activity helps students realize that "disease is a breakdown in structures or functions of an organism … [in this case as a] result of damage by infection by other organisms" (NRC 1996, p. 155).

Grades K–3: Microbe Mania

Purpose

Students will explore microbes (yeast) by growing them on an agar medium.

Note: Environmental bacteria should never be cultured in a classroom.

Materials

- Active dry yeast
- Sterilized petri dishes
- Agar medium with glucose
- Warm water
- Magnifying lenses

Procedure

1. Read *Germs Are Not for Sharing* to the class and point out that germs are everywhere—in the air, on food, in the water. Ask students how they try to avoid germs since they are all around us. Some possible answers include washing hands, not putting hands or objects in your mouth, and covering your mouth when you cough." One of the points made in the book is that not all "germs" are bad.

2. Introduce the idea to the students that they can grow their own microbes (which in this case will be yeast). All they need is some type of food to grow on and moisture.

3. Begin by having the students prepare the food medium (in this case, ask your high school biology teacher to prepare petri dishes with an agar that has glucose [sugar] in it to allow yeast to grow) by placing a few drops of warm water on top of the agar.

4. Have students sprinkle active dry yeast (not instant) on top of the agar in the petri dishes so that a light film covers the dish. Active dry yeast is still a living organism and needs to be reactivated. The warm water that was placed on top of the agar will reactivate the yeast. Tape each

Disease Detectives

petri dish closed after adding the yeast. Other airborne microbes can land and take root on the agar, so taping the petri dish closed minimizes student contact should this occur. Keep an eye out for mold, which appears furry, and discard the petri dish immediately if it appears. Have students record observations.

5. Have the students place the samples in a warm, dark place (light can kill some microbes and cold stunts the growth of microbes) for two days. Then, have the students make observations about what they see using magnifying lenses. Questions that can be posed to the student at this time include "What do you notice about the samples?" Students may respond with, "Things are growing on the agar," which leads into a discussion about how we placed the yeast on the agar to allow it to grow. Students may notice that the yeast multiplied or got bigger compared to the original amount. This connects to the idea that any microbe multiplies under the right conditions. Ask, "Where do you think microbes that we didn't put in food or on the agar came from?" (the air or our hands). Introduce the idea to the students that germs or microbes can be floating in the air even though we may not see them because they are microscopic. An example is the cold virus, which can be spread through coughing into the air (the reason we have students cover their mouths when they cough).

6. Repeat looking at the samples every day for a week and ask the students to continue to make observations regarding the growth of the yeast. You may want to connect this to the idea that microbes can land on food when we don't know it and grow if given the proper conditions. This is why bread that is old and left on the counter (even in its wrapper) can begin to get moldy or why sometimes food begins to smell when it starts to go bad.

7. When all observations are complete, dispose of samples properly or return glass petri dishes to

your biology teacher for proper disposal. Students should be reminded to always wash their hands following the handling of any containers.

8. Return to the book to have students consider all of the places that germs can be located and ways to prevent the spread of germs.

Grades 4–6: Catching a "Cold"

Purpose

Students will participate in a simulated epidemic where they trace the spread of the epidemic to the "zeroth" patient (initiation point).

Materials

- Glo Germ Powder (see Internet Resources)
- Portable ultraviolet light
- Catching a Cold Record Sheet (p. 227; one per student)
- Safety goggles

Procedure

1. Using *What Are Germs?* read to the students over several days, each time following with discussion. When you get to the part related to "harmful germs," ask students what *infectious* means and whether they have ever had any common infectious diseases.

2. Ask the class, "What happens when you 'catch' a cold?" Then ask students, "How is the cold virus transmitted between humans?" Some have a hard time understanding that humans can spread a disease, in this case, a cold, to one another.

3. Gather other student input on what other types of diseases they have heard of and how they are spread. Explain to the students that they will participate in a simulation, which is a chance for students to represent or model something

that happens in real life in a classroom setting. Review safety considerations for using any unknown substance in a classroom, which include wearing safety goggles and not touching your hands to your eyes, mouth, or nose.

4. Ahead of time solicit the help of one student who can play along and not give away that he or she is the carrier of the cold virus. Alternatively, the teacher can serve as the carrier. This student should liberally sprinkle Glo Germ Powder onto the fingers and palm of the hand that he or she shakes with—usually the right hand.

5. Remind students that germs or viruses can be spread through casual contact. Students will then shake hands with someone else in the class and record it on the Catching a Cold Record Sheet. Continue this process through three to four different cycles, making sure the shaking of hands happens with a different person each time and is recorded.

6. Ask students to hold their hands palm upward on their laps. Dim the lights in the room. Circulate around the room and shine an ultraviolet light onto each person's extended hands. If a person came in contact with the initially infected carrier his or her hands will glow under the ultraviolet light. Though not considered

hazardous when used correctly, UV-A rays are associated with health risks such as skin cancer and cataracts. Be aware of these risks when deciding to use UV-A lights in the classroom and always follow manufacturer's instructions.

7. Explain to the students that any glow on the hands indicates a positive result for the "germ," or in this case "cold virus." If they had been coughed on or shaken hands with a person who really did have a cold, they could have been exposed to the cold virus and could get sick.

8. Ask the students to use their record sheets to determine who the initial "sick" kid was.

Although no one likes to get sick and pass along germs, these activities help the students see that their actions can help them keep healthy.

Reference

National Research Council (NRC). 1996. *National science education standards.* Washington, DC: National Academies Press.

Internet Resources

Using Glo Germ
www.glogerm.com/using.htm

Glo Germ
www.teachersource.com/BiologyLifeScience/ Germs/GloGerm.aspx

Disease Detectives

Catching a Cold
Record Sheet

Trial #1.

_____ shook my hand and I shook _____'s hand. After shaking
hands, I observed my hand and these are my observations: _____
_____ .

Trial #2.

_____ shook my hand and I shook _____'s hand. After shaking
hands, I observed my hand and these are my observations: _____
_____ .

Trial #3.

_____shook my hand and I shook _____'s hand. After shaking
hands, I observed my hand and these are my observations: _____
_____ .

Trial #4.

_____ shook my hand and I shook _____'s hand. After shaking
hands, I observed my hand and these are my observations: _____
_____ .

When my teacher shined an ultraviolet light on my hands, they _____ (did/did not)
glow.

This means that I _____ ,
and if the powder was the cold virus I _____ .

From our discussions, the person who had the original cold virus was
_____ because _____
_____ .

Chapter 41

Rock Solid Science

By Karen Ansberry and Emily Morgan

Children are naturally curious about the world around them, including the rocks beneath their feet. By observing, describing, and sorting a variety of rocks, students can discover that rocks have certain physical properties by which they can be classified. The activities described here take a fun approach to learning about the properties, uses, and formation of rocks. Students in grades K–3 explore the properties and uses of their own "pet rock," while students in grades 4–6 investigate "rock stories."

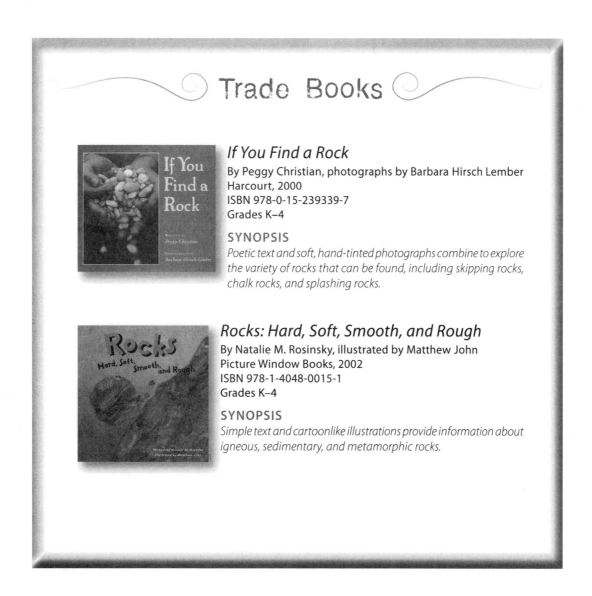

Trade Books

If You Find a Rock

By Peggy Christian, photographs by Barbara Hirsch Lember
Harcourt, 2000
ISBN 978-0-15-239339-7
Grades K–4

SYNOPSIS

Poetic text and soft, hand-tinted photographs combine to explore the variety of rocks that can be found, including skipping rocks, chalk rocks, and splashing rocks.

Rocks: Hard, Soft, Smooth, and Rough

By Natalie M. Rosinsky, illustrated by Matthew John
Picture Window Books, 2002
ISBN 978-1-4048-0015-1
Grades K–4

SYNOPSIS

Simple text and cartoonlike illustrations provide information about igneous, sedimentary, and metamorphic rocks.

Rock Solid Science

Curricular Connections

The *National Science Education Standards* state that students in grades K–4 should understand that earth materials include solid rocks and soils, and that these materials have different physical and chemical properties that make them useful in different ways (NRC 1996). The Standards also suggest that young children be encouraged to closely observe the objects and materials in their environment, note their properties, and distinguish them from one another.

Following these suggestions, the K–3 lesson focuses primarily on recognizing properties of rocks (i.e., shape, size, color, texture, and luster), understanding how properties of rocks can be used to sort them, and exploring how a rock's properties and its uses are related. The Standards advise that for grades K–4, the study of rocks not be extended to the changes in the solid earth known as the "rock cycle" because this concept has little meaning to young children. The Standards do suggest that students in grades 5–8 develop an understanding of the process by which old rocks at the Earth's surface weather, forming sediments that are buried, then compacted, heated, and often recrystallized into new rock. Eventually, those new rocks may be brought to the surface and the rock cycle continues. The 4–6 lesson is based on the idea that every rock has a "story" that can be uncovered through observation and research. Students observe different rock samples, discover that rocks are composed of minerals, and learn that rocks can be classified as igneous, sedimentary, or metamorphic depending upon how they are formed. They form "rock groups" to do research on a particular rock, then create a picture book to tell the story of that rock.

Grades K–3: Pet Rocks

Materials

- "Found" rocks
- Sticky notes
- Poster paper
- Markers or crayons

Engage

Introduce the book *If You Find a Rock*. Build connections to the reading by telling students that the author is a "rock hound" (a person who loves to collect rocks), and ask if any of them would consider themselves to be a rock hound. Explain that while you are reading the book aloud, you want them to think about what some of the rocks are used for and what properties, or characteristics, make them suited for that use. After reading, discuss the various uses and properties of the rocks in the book. For example, a skipping rock is used for skipping across water. The properties that make it suited for that purpose are its flat and round shape and its small size. Shape and size are properties of rocks. A chalk rock is used to make pictures on the pavement. The properties that make it suited for that purpose are its white color and its soft, dusty texture. Color and texture are also properties of rocks. Another property of rocks is luster, or how the minerals in rocks reflect light. Words that describe a rock's luster include shiny, dull, and sparkly.

Explore/Explain

Ask students if they have ever heard of a "Pet Rock." Explain that way back in 1975, a businessman in California came up with the idea of selling rocks as pets. The Pet Rock became a huge hit. Tell students that they are going to be rock hounds on the hunt for their own pet rocks. They can go home and search for rocks with adult supervision or select rocks from their own collections. Discuss these rules: Your pet rock must *be smaller than a tennis ball*. You are not allowed to throw your pet rock. The next day, give students rulers and hand lenses to observe their pet rocks' properties, including shape, size, color, texture, and luster. Discuss how observations such as *big* or *small* are not scientific observations because they are not exact. Using measurements to describe the size of a rock is more scientific. Demonstrate how to determine the length of a rock by measuring its longest side in centimeters. Have students choose three properties of their rocks and write each property on a separate sticky note. Next, make a Properties Chart on the board with six columns labeled Shape, Size, Color, Texture, Luster, and Other. Model how to use the chart by having one student share a sticky-note observation and then place it in the appropriate column on the chart (e.g., smooth goes in the texture column). Then have small groups of students take turns placing all of their sticky notes in the appropriate columns on the chart. Some students may need help in determining which observations go in which columns.

Elaborate

Have all the students stand, holding their pet rocks. Choose one sticky-note observation from the Properties Chart, such as smooth. Students will then determine whether or not their rocks are smooth, and form two groups in the room: smooth and not smooth. Have them compare their rock to the others within their group, and then switch groups if they wish. Repeat this sorting and classifying process using several more observations from the Properties Chart. Explain that scientists use properties like color and texture to help them classify rocks.

Evaluate

Younger primary students can draw detailed pictures of their rocks, listing as many properties as they can. For older primary students, explain that the original Pet Rock was packaged in a box that looked like a pet carrying case. The Pet Rock inventor used creative advertisement to help sell his product.

Have students create advertisements for their pet rocks. The ads should show what they have learned about properties of rocks, including

- descriptions of the rock's shape, size (including measurements), color, texture, luster, and any other properties;
- suggested uses for the rock based upon its properties; and

- labeled drawings of the rock highlighting its unique features.

For fun, ads can also include

- drawings of the rock's packaging (crate, box, bag, etc.);
- training tips for the pet rocks; or
- commercials, slogans, or jingles.

Afterward, have students share their advertisements with the rest of the class.

Grades 4–6: Rock Stories

Materials

- Hand lens
- Centimeter ruler
- Rocks
  * Obsidian
  * Granite
  * Sandstone
  * Limestone
  * Marble
- Balance or scale
- Supplies for creating picture books
  * Paper
  * Crayons or markers
  * Stapler or brads
- Rock Properties student page (p. 233)

Engage

With a rock hidden in your hand, announce that you are holding something that is older than them, older than the school building, even older than you … something that could even be millions of years old! Have students guess what it is. Reveal the rock, and then tell students that a rock is probably the oldest thing they will ever touch. Explain that every rock has a "story" that they can uncover through careful observation and research … a story about how it formed, what it is made of, and how it can be used.

Rock Solid Science

Explore/Explain

Give each student a hand lens, centimeter ruler, and one of the following rocks: obsidian, granite, sandstone, limestone, or marble. Tell them that to uncover the rocks' stories, they can begin by observing physical properties including shape, size, color, texture, and luster. Model how to make good qualitative and quantitative observations of a rock's properties. For example, "My rock is irregular in shape, has a mass of 56 g, and is 8 cm long. It is reddish brown, with a dull luster and rough texture. It has grains the size of sand." Students should measure their rocks in as many ways as they can (length, circumference, mass, etc.). Have students use the Rock Properties student page to record their observations and then share with others.

Next, tell students that the picture book *Rocks: Hard, Soft, Smooth, and Rough* can give them more clues about their rocks' stories. Each one of the rocks they have been observing is described in the book. Have them signal when they hear a description that matches their rocks as you read the book aloud. After reading, discuss why properties such as color, texture, and luster might be better ways to identify rocks than size or shape. Then have students use hand lenses to look for specks, crystals, grains, or stripes in their rock samples. These are the minerals that make up their rocks. Some rocks are made of a single mineral, but most are made of several minerals.

Elaborate/Evaluate

After all students have identified their rocks, have them form "rock groups" with other students that have the same rock. These groups can use books and websites to research their rocks' stories to create picture books. Each book should include the name of the rock, its properties, the minerals it is composed of, the rock group to which it belongs (igneous, metamorphic, or sedimentary), an explanation of how it formed, and some possible uses. For fun, groups can even write rock songs! Have each group share its picture book with the rest of the class or with younger students.

Reference

National Research Council (NRC). 1996. *National science education standards*. Washington, DC: National Academies Press.

Rock Hound's Name: _____

Rock Properties

SHAPE How would you describe its shape?	COLOR What colors or patterns does it have?	TEXTURE How does it feel?	LUSTER How shiny or dull is it?	SIZE What is the longest length in cm?

OTHER OBSERVATIONS	LABELED DRAWINGS

Chapter 42

Rocking Around the Rock Cycle

By Christine Anne Royce

"What type of rock is this?" is a common question spoken by all children at some point in their young lives. Many students have rocks of all shapes and sizes in their "collections." With these activities, young students will have the opportunity to explore more in depth as they make observations of the three types of rocks—igneous, metamorphic, and sedimentary. Upper-grade students who have previously studied rocks can use these books to review what they know of the rock cycle.

Trade Books

Let's Go Rock Collecting
By Roma Gans, illustrated by Holly Keller
Harper Collins, 1997
ISBN 978-0-06-445170-3
Grades preK–3

SYNOPSIS
Children go on a rock collecting expedition. This Let's Read and Find Out About Science series book introduces students to the different types of rocks and how they are formed, and it builds excitement about collecting rocks. The book can be a useful resource for helping examine different types of rocks and what happens to them over time. It should be noted that they use the term rocks loosely in this book and include some minerals. The discussion of the hardness scale is for minerals only.

The Rock Factory: A Story About the Rock Cycle
By Jacqui Bailey, illustrated by
Matthew Lilly
Picture Window Books, 2006
ISBN 978-1-4048-1596-4
Grades 2–5

SYNOPSIS
The author uses colorful pictures and creative text to describe the rock cycle, differences between rocks and minerals, and other related topics. This book can be used to assist the reader in understanding the different types of rocks and how they are formed.

Rocking Around the Rock Cycle

Curricular Connections

Who hasn't had a rock collection at some point in their lives? It may have been rocks that were a certain color or certain shape, or rocks that sparkled. Regardless, rocks are all around us—in streams, on paths, in fields—they are literally beneath our feet every day. By using *Let's Go Rock Collecting* as a springboard, teachers can begin to ask students what they know about rocks—"Where did they come from? What are they made from? How have they changed over time?" All of these questions can be posed to the student throughout the reading of this book. Young students will generate many similar questions—for example, "Why are they different colors?" and "Why do some have stripes and others have spots?"

This excitement about such a common object can be used to meet part of the *National Science Education Standards* and have students make observations about and describe the properties of rocks. At this age, the use of the science-process skills of observation and classification into groups and the use of language skills for providing detail about rocks is sufficient. The NSES point out that "understanding rocks and minerals should not be extended to the study of the source of the rocks, such as sedimentary, igneous, and metamorphic, because the origin of rocks and minerals has little meaning to young children" (NRC 1996, p. 130). An introduction to rock collecting helps students begin to understand that each rock has its own unique features.

Older students can then explore how and where individual rocks fit into a larger process—the rock cycle. Students at this level have likely learned about the geological changes that happen as different types of rocks—igneous, sedimentary, and metamorphic—form through the rock cycle. *The Rock Factory* serves as a good resource to review these concepts as they complete an assessment activity that demonstrates their understanding of the different ways rocks form.

Grades K–3: Examining Rocks

Purpose

To examine and make observations about the three different types of rocks—igneous, sedimentary, and metamorphic

Materials

*   A hand lens (one per student)
*   Rock samples for each group (Samples can be borrowed from a local high school or university or purchased through a science supply company.)
    *   Granite
    *   Basalt
    *   Obsidian
    *   Sandstone
    *   Shale
    *   Conglomerate
    *   Marble
    *   Gneiss
    *   Limestone
    *   Pumice or scoria
    *   Quartzite
*   A science notebook or graphic organizer for each student
*   Rock Observation Chart (p. 239)

Procedure

1. While reading *Let's Go Rock Collecting* to the students, ask "What are the children collecting?" "Are all the rocks they talk about all the same?" or "How do the children describe the rocks they find?" to help students construct an information sheet about rocks. Have the students describe what a particular rock in the book looks like.

2. Distribute the rock samples to each group and have students record observations about the rock properties (e.g., texture, appearance) on a chart. While working in groups, ask the

Rocking Around the Rock Cycle

students to come up with a way to sort or classify the rocks into groups. After groups of students make observations and determine a classification system for their rocks, ask them to share with the class why they selected the groups they did and what were the common properties the rocks shared. Developing their own classification system allows students to begin to recognize the concept of properties. Geologists use properties such as texture, grain size, crystal size, types of layering, and so on to identify the different types of rocks, and students will be making some initial observations about some of these properties in this activity.

3. Introduce the idea that students can identify the general type of rocks they have by making observations of their appearance. Distribute a sample of the same type of granite to each group along with the Rock Observation Chart (p. 239).

4. After students describe their observations, the teacher can begin to introduce vocabulary words. Some of the words might include crystals, which are sometimes observable in igneous or metamorphic rocks, and layers, which often appear in sedimentary rocks. The actual identification of all rocks is more complicated than this initial task, but students can refer to *Let's Go Rock Collecting* to reorganize their groups into the three main types of rocks—igneous, sedimentary, and metamorphic—based on their properties. (The rocks named above would fall into the following categories: igneous—granite, basalt, obsidian, pumice or scoria; sedimentary—sandstone, shale, conglomerate, limestone; metamorphic—marble, gneiss, and quartzite.) The book does a nice job of providing a picture of some of these rocks along with a description of each.

5. Understanding that there are different types of rocks prepares students to later explore how the three different types of rocks form.

Grades 4–6: Completing the Cycle

Purpose

To assess students' understanding of the rock cycle through a creative, hands-on assessment that allows students to assemble a model of the rock cycle

Materials

- Craft materials such as poster board, construction paper, glue, scissors, tape, markers, and crayons
- Reference books (see Resources)
- Rock samples
  * Granite
  * Basalt
  * Obsidian
  * Sandstone
  * Shale
  * Conglomerate
  * Marble
  * Gneiss
  * Limestone
  * Pumice or scoria
  * Quartzite

Procedure

1. Following a unit on rocks in which students learned about the rock cycle, students conduct this activity to help synthesize their learning. After reading the book *The Rock Factory,* review with students that the three main types of rocks form under different circumstances—igneous rocks form when molten magma or lava cools; sedimentary rocks form from the layering and compaction of sediments; and metamorphic rocks form from intense heat or pressure. It should be noted that it is not a circular cycle, as

Rocking Around the Rock Cycle

different rocks can get caught in a smaller part of the cycle. In other words, not all igneous rocks become sedimentary, nor do sedimentary in turn become metamorphic. A great graphic of this can be found at *www.personal.psu.edu/users/c/l/cll161/insys%20441/main.html*.

2. Using either samples of actual rocks or pictures of the rocks, ask the students to attempt to match an igneous or sedimentary rock with the rock it can be changed into after the rock has gone through metamorphism. Examples of sequencings could be granite to gneiss, sandstone to quartzite, limestone to marble, and shale to slate. Although this may be difficult, hopefully students will be able to see that some metamorphic rocks have similar characteristics of the rocks from which they came. For example, some limestone rocks are creamy white or pale grey in color and appear to be layered and flat, whereas marble is a similar color. It is important for teachers to match up the "before" and "after" rocks carefully to aid the student in seeing similarities. Have students make statements about why they think the "changed" rock was at one point the rock they chose. It could be something like, "Gneiss has the same color minerals in it and the spots look like they are changed somehow" or, "Sandstone looks like it has tiny crystals or pieces of sand cemented together, and quartzite has tiny crystals as well." When the students describe their observations, have them try to describe how the rocks may have formed as well. For example, metamorphic rocks undergo intense heat and pressure to change the original rock into a new rock, but the new rock still comprises the same minerals as the original rock. To help with this task, create a list of key vocabulary on the board (e.g., igneous, sedimentary, metamorphic, weathering, compaction, magma, cooling, cementation, and deposition).

3. Using *The Rock Factory* and other references, ask students to construct a model of the rock cycle. They may want to simply draw it, or they may want to find and cut out pictures of the different types of rocks for each part. While they are doing this, they should be using the key vocabulary terms identified during the reading.

4. When students have completed their models of the rock cycle, they can share them with their group or the class.

Regardless of whether children are making observations about rocks and developing an initial idea about what a rock is or assembling a model of the rock cycle to serve as an assessment of their understanding, it is important to allow them to engage in activities that definitely ROCK!

Reference

National Research Council (NRC). 1996. *National science education standards.* Washington, DC: National Academies Press.

Resources

DK Publishing. 2003. *Rocks and minerals.* New York: Author.

Staedter, T. 1999. *Reader's digest pathfinders: Rocks and minerals.* Pleasantville, NY: Reader's Digest Books for Children.

Rock Observation Chart

ROCK SAMPLE	APPEARANCE	TEXTURE	MORE THAN ONE SUBSTANCE?	OBSERVATIONS
Granite	The rock looks jagged and has different colors.	Rough, nothing comes off in your hand.	Yes, there are three types of spots—white, pink, and black.	The rock looks spotted or speckled.

Chapter 43

The Dirt on Soil

By Christine Anne Royce

Have you ever played in the dirt? Repotted a plant? Planted a garden? In each and every one of these activities, you touched something that is a valuable resource in our lives—dirt! In these activities, students have a chance to investigate different aspects of something that is right beneath our feet but plays an important role in the daily lives of many organisms.

Trade Books

Dirt

By Steve Tomecek, illustrated by Nancy Woodman
National Geographic, 2002
ISBN 978-0-7922-8204-4
Grades preK–3

SYNOPSIS

A personable mole takes the reader on a tour of dirt and introduces the different creatures that live in it. Throughout this tour, the reader is introduced to how soil is formed, why it is important to living things—plants and animals—and the different types of soil one might find.

A Handful of Dirt

By Raymond Bial
Walker and Company, 2000
ISBN 978-0-8027-8698-2
Grades 3–6

SYNOPSIS

A Handful of Dirt *explains what components comprise soil and describes the important role that soil plays in keeping things alive—from microbes to moles. The author includes basic instructions for setting up a home compost heap.*

Curricular Connections

"Don't play in the dirt!" If parents had a nickel for every time … well, you get the idea. Whether it is the feel of the moist soil in their hands or the grains of sand running between their fingers, students are drawn to this substance. *Dirt* and *A Handful of Dirt* both approach the topic from the perspective that soil is a resource many different organisms rely on—and one that we should protect. Each book also provides information on the structure of soil and how it forms over time.

These aspects can help bridge different subject areas of the *National Science Education Standards*. In addition to being part of the Earth sciences area, the topic of soil shows the interconnectedness of different subjects—bringing in life science and ecology and meeting the goal that "concepts and processes provide connections between and among traditional scientific disciplines" (NRC 1996, p. 115). Students will love jumping in and getting their hands dirty once they have a better understanding of what goes into the production of soil and why it is one of our most valuable resources.

Grades K–3: Seeing Soils

Purpose

To make observations about different types of soils.

Materials *(per group of three to four students)*

- Samples of clay, silt, sand, and gravel (purchased at garden supply or home improvement stores)
- Four different soil type samples (e.g., wooded/leafy area, bank of a stream or pond, drier/sandy area, grassy area)
- Hand lens
- Chart paper
- Group K-W-L chart
- Markers
- The Dirt on Soil K-W-L chart (p. 245)

Procedure

1. Show students the cover of *Dirt* and ask them what they already know about dirt. Have the students fill in the first column of the Dirt on Soil K-W-L chart (p. 245). Student responses might include, "Things crawl in the dirt." Ask them to complete the "want to know" column while you guide them to focus on questions that can be answered through observing the samples.

2. After students generate lists of questions they want to know, read *Dirt* by Steve Tomecek to them. Have them pay particular attention to the fact that soil comes in different colors and textures and that the *sediment* (the materials that form soil) in soil can come in different sizes.

3. Provide each group of three to four students with a small container of each of the following sediment types—clay, silt, sand, gravel—without telling them the name of each substance. Ask the students to make observations about each of the different types of sediment types with a hand lens and develop a description for each type, such as "clumps together" or "looks like small rocks."

4. Have the groups share their observations with the entire class. Refer back to page 9 of the book, and ask the students to identify the four different materials based on the descriptions in the text (e.g., "Grains of sand are about the size of grains of sugar").

5. Give the student groups samples of different soils at this point (do not use potting soil from the store, as it contains components meant to retain water and doesn't contain many of the organic materials found in natural soils). Have them repeat the process of making observations of the soil types. Do they see any different types of sediment? Do some soils hold or retain more water than other soil types?

The Dirt on Soil

6. Discuss other things found in soil, such as the different types of organic material identified on page 13. Ask students where twigs, dead leaves, and other organic materials might come from (from decaying plants and animals).

7. After the students have had a chance to investigate the different types of sediment and soils, have them return to the K-W-L chart they started and ask them to review the *K* column and correct any item that might be in error, then complete the *L* column with what they learned about soils. Possible answers include, "Soil is made up of different size sediments and organic materials."

Soil Safety

Follow these soil safety guidelines

1. Know the source of your soil samples! Soil can be contaminated by pesticides, animal waste, and so forth.

2. Obtain parent/guardian permission before having students work in soil or in compost to inform them of possible allergens (e.g., mold/spores), which might affect students with compromised immune systems, allergies, or asthma.

3. Have students wear plastic gloves and make sure all open cuts or scratches are covered minimally to prevent infection, and always wash hands with soap and water after working with soil or compost. Wash desktops where soil activities took place with mild soap and water. Do not allow snacks or other food products during soil activities. Don't keep wet soil more than a day or two. Mold and bacteria spores will grow in it.

4. Wear appropriate clothing (long sleeves and pants) and closed-toe shoes or sneakers when working in a compost pile.

5. Handle compost materials, wire mesh, stakes, and wooden boards with care and caution. Use only nonmercury thermometers.

Grades 4–6: Creating Compost

Purpose

Students create a school-yard compost heap to investigate the effects different organisms have on soil.

Materials

- Compost or soil for each group of students
- Wire mesh
- Stakes
- "Brown stuff" and "green stuff" (see descriptions below)
- Additional compost or soil for the bin
- Water
- Magnifying lenses
- Thermometers

Procedure

Note: *Local/State agricultural offices often have information on composting and perhaps offer composting bins.*

To begin this long-term activity of making a school yard compost pile, coordinate a location with your school office—it should be a shady 4 ft. by 4 ft. spot that will not be disturbed. Enclose your area with either wooden boards or wire mesh, as described on pp. 26–27 of *A Handful of Dirt*. See "Soil Safety" for guidelines.

1. After reading *A Handful of Dirt*, give each group of students a sample of fresh soil or compost and ask the students to recall how soil is made

from the book. Where do the different components come from? (Organic materials in the soil come from decaying animals and plants, and the inorganic materials come from weathered rocks.) Revisit page 27, in which students are provided with basic information about how to create a school-yard compost heap.

2. Ask students to collect the materials for creating the compost—"green stuff" (grass clippings, weeds, kitchen vegetable scraps, flowers, etc.) and "brown stuff" (dead, dried plant parts such as leaves, twigs, and shredded newspaper). Do not add milk products, meat, fats, oils, or juices to the mixture as they may attract animals. You will also need water, air, and microorganisms (found in additional compost) as the other components.

3. After constructing your enclosure, have the students alternate layers of brown stuff (high in carbon) and green stuff (high in nitrogen), beginning with a layer of dead leaves 15–25 cm thick. Follow this with a layer of green stuff of the same thickness, mixing the layers slightly as you build your layers. Add some compost or soil (not potting soil) as the next layer—this intro-

duces microorganisms to the mixture. Finally, sprinkle the compost bin with a water—the material should be damp like a sponge but not soggy. Repeat this procedure of layering until the bin has two to three layers.

4. Ask students to explain what each component might be if this was a natural setting such as a forest (e.g., dead plants, water from precipitation).

5. Maintaining the compost heap (turning the layers occasionally and adding more components) teaches students about decomposition and is a worthy recycling project. In addition, students can take temperature readings; observe changes over time and when different components are added or the materials are mixed; connect the pile to the process of soil formation; and discuss how the compost pile represents an ecosystem in itself.

Reference

National Research Council (NRC). 1996. *National science education standards*. Washington, DC: National Academies Press.

The Dirt on Soil

K	W	L
(What I Know)	(What I Want to Know)	(What I Learned)

Chapter 44

Fascinating Fossil Finds

By Christine Anne Royce

Dinosaurs and the prehistoric age captivate children. At a young age, children can often cite facts and information about various dinosaurs, even if they do not yet understand the connection between dinosaurs and the way scientists gather information from fossil evidence. Through these activities, students "unearth" fossils and explore the processes scientists use in uncovering these fascinating finds and interpreting Earth's past.

Trade Books

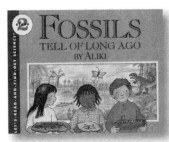

Fossils Tell of Long Ago
By Aliki
HarperCollins, 1990
ISBN 978-0-06-445093-5
Grades K–3

SYNOPSIS

The author depicts how scientists uncover, preserve, and study fossils. Information about what fossils tell us about long ago is also provided.

Dragon in the Rocks: A Story Based on the Childhood of the Early Paleontologist Mary Anning
By Marie Day
Owl Communications, 1992
ISBN 978-0-920775-76-9
Grades 2–5

SYNOPSIS

This fictionalized trip through the childhood of Mary Anning describes how she supported her family by finding and selling fossils in Lyme Regis, England. The story also highlights the fact that Anning is credited for finding and preparing the first Ichthyosaurus known to England and the first Plesiosaur. What started out as a young child's interest grew to become an important contribution to the scientific community.

Curricular Connections

Dinosaurs and fossils are often found in the elementary curriculum. Teachers and students alike work their way through interpreting the names of different dinosaurs and examining different kinds of fossils, in addition to spending many hours reading about dinosaurs. The *National Science Education Standards* support this student-generated interest in the study of fossils at the K–4 level, stating, "fossils provide evidence about the plants and animals that lived long ago and the nature of the environment at that time" (NRC 1996, p. 124).

Students in grades K–4 should be able to use data to construct a reasonable explanation as well as "learn what constitutes evidence and judge the merits or strength of the data and information that will be used to make explanations" (NRC 1996, p.122). This ties in to the *National Science Education Standards* for grades 5–8 that asks students to think critically and logically to make the relationships between evidence and explanations. Older students can begin to examine how "fossils provide important evidence of how life and environmental conditions changed" (NRC 1996, p. 160). For example, students can talk about how areas that in ancient times were wet and forested and had flourishing plant life are now dry and desertlike areas.

Another important but often-neglected area in the science curriculum is the history and nature of science. *Dragon in the Rocks*, though a fictionalized account, is a wonderful book that examines how a young girl took fossils and assembled them into the prehistoric animal they represented. The ability to construct a fossil of a now extinct animal in itself is important for young children to learn because understanding science involves not only the science content (e.g., names of fossils, dinosaurs) but also understanding the nature of science and processes that scientists use to develop theories.

Grades K–3: Creating a Cast and Mold

Purpose

In this activity, students learn about fossil formation as they create replicas of *mold* and *cast* fossils.

Materials

*   Seashells, twigs, or small rocks (the more detail the better)
*   Petroleum jelly
*   Plaster of paris
*   Measuring cup
*   Small margarine containers or other plastic dishes

Procedure

1. Begin by discussing with the students what a fossil is and how scientists use fossils to help explain the past. Ask, "Have you ever seen a real dinosaur?" or, "When you visit a museum, have you ever seen a skeleton of a dinosaur?" While discussing these questions refer to the different pages in *Fossils Tell of Long Ago* to allow students to hear the information and gather visual clues from the pictures. Common misconceptions should be clarified: There are currently no living dinosaurs, and humans and dinosaurs never coexisted. So, the only evidence we have about dinosaurs is what can be inferred from fossilized remains. Almost all fossils are found in sedimentary rock. Explain to the students that the plaster of paris represents the soft sediment that an object would fall into before it becomes a fossil.

2. Ask the students to select either a shell, twig, or rock from which to make their "fossil." Have the students coat their object with a light layer of petroleum jelly (to prevent the plaster of paris from sticking to it). Students should put approximately one-half cup of plaster of paris in the bottom of their margarine container and

gradually add water until the plaster of paris is thick (the ratio of water to plaster of paris will be about one part water to two parts plaster of paris). Allow the plaster of paris to sit for one to two minutes until it starts to harden. At this time, students should place their objects coated with petroleum jelly into the plaster—but not submerge it—and allow it to dry overnight.

3. The next day, have students carefully remove their objects from the plaster. Introduce the word *mold* to the students and explain that this is one type of fossil that helps scientists gather evidence about the past. Molds are an imprint of the object. Allow the students an opportunity to examine the detail in the plaster of paris as well as to compare its similarities (shape, pattern, etc.) to the original object. After students have had an opportunity to explore for a short time, introduce the word *cast* to them. This is a replica of the original fossil in a different material (in this case, plaster of paris) and can form when a mold is later filled in with sediment.

4. Have the students coat the entire surface of the dried plaster mold with a thin layer of petroleum jelly to prevent sticking. In a separate container, have each student mix up plaster of paris as before. Carefully pour the plaster of paris into a container on top of the mold and surrounding material. Explain that this layer of plaster of paris represents additional sediments that buried the original item. (Note: In a real situation, the original material would either decay or be dissolved by groundwater, resulting in no original material remaining.) Allow the cast to dry overnight. When the plaster is set, carefully separate the cast (top part) from the mold (bottom part) along the layer of petroleum jelly. Once again, allow students to examine and make observations of the cast, comparing it to the original object.

5. Students can switch molds and casts with other students to try and identify the object that made the fossil. While they are engaging in guessing what created the fossil, the teacher can draw them back to how scientists do not know but rather make observations and inferences about fossils that are found.

Grades 4–6: Excavating Evidence

Purpose
This activity places students in the roles of paleontologists excavating a fossil bed.

Materials
- Shirt box or shallow plastic pan (one per group)
- Plaster of paris
- Sand
- Water
- Assortment of "fossils" (e.g., shells, twigs, rocks, and pebbles)
- Safety goggles
- Excavating tools
  * Plastic spoons
  * Paper clips
  * Toothpicks
  * Paintbrushes
- Fascinating Fossil Finds student page (p. 251)

Procedure
1. Scientists make careful observations and measurements and record data when they "discover" an area in which fossils are buried. In *Dragon in the Rocks*, students read about Mary Anning chipping away at the rocks to free the bones she found with the intent to reassemble her "great creature." The line, "She had drawn a picture of the whole skeleton as best she could and had given a number to each bone," shows how scien-

tists record information to develop a complete picture of the evidence gathered.

2. This activity is preparation-intensive but provides an excellent opportunity for students to take on the roles of scientists. First, develop a "fossil bed" for each group of four to six students. This fossil bed can be created using either a shirt box or shallow plastic pan. Mix two cups of plaster of paris with one cup of sand and water until it reaches a thick consistency and spread the mixture in the bottom of the shirt box to a depth of 5 cm. Place various different "fossils" and fossil fragments in the bed, such as broken up shells, twigs, textured rocks, or pebbles. (A real fossil excavation site would have a mix of many different types of fossils, but for simplicity you can use a separate type of fossil for each bed.) *Partially* submerge each item in the plaster, so students can "excavate" the fossils. Once the plaster of paris is set, cover the bed with additional sand to "bury" the fossils. Using a string, mark off a grid on the box that is three squares by five squares.

3. Introduce the activity by explaining that the class has stumbled upon a fossil bed and needs to unearth the fossils and gather evidence as to what the site may have been like millions of years ago. Students should be shown how paleontologists mark off sites using standard-size grid squares and sketch what is found in relation to the grid squares to preserve the visual evidence before the fossils are removed.

4. When students are ready to excavate, have them put on safety goggles to prevent spraying sand plaster from getting in their eyes. Then distribute plastic spoons to remove some of the sand on top and start the excavation. As they get to the fossils, distribute paper clips, toothpicks, and paint brushes for students to use to preserve the fossils in the best shape that they can.

5. When students find a fossil, they should make field notes about the measurements of the object, the position in which they find it, and its location in relation to other fossils on the Fascinating Fossil Finds student page. Students should use as much detail and precision as possible when sketching their finds on the grid. In addition to observations, students should make and record any inferences they develop based on their findings (e.g., a shell may indicate a marine past). Students should remove the fossil and number it, which will allow them to reassemble their fossil evidence at a later date.

6. When students have completed the excavation, discuss how different fossils are found at different depths in real digs, representing a difference in time (i.e., the deeper the fossil the older it is because that sediment was layered first).

7. After the different fossils have been unearthed, students can assemble them, explain what the fossil represents, and justify why, based on the evidence, they arrived at their theory, just like real paleontologists.

Reference

National Research Council (NRC). 1996. *National science education standards.* Washington, DC: National Academies Press.

Fascinating Fossil Finds

Number and indicate what type of fossilized evidence you found in what square.

	1	2	3
1			
2			
3			
4			
5			

Chapter 45

Mysteries of the Past

By Emily Morgan, Karen Ansberry, and Katie Davis

Paleontologists, scientists who study the history of life on Earth, work in a dynamic area of science. Think of putting together a jigsaw puzzle with most of the pieces missing—that's what creating the fossil record is like. Each time a new piece is discovered, ideas about the whole picture become clearer (and sometimes even change considerably). These activities focus on how we know what we know about prehistoric life, how scientists' ideas have changed over time, and what mysteries of the past remain to be solved.

Trade Books

Prehistoric Actual Size

By Steve Jenkins
Houghton Mifflin, 2005
ISBN 978-0-618-53578-1
Grades K–3

SYNOPSIS

Jenkins's trademark paper collage illustrations depict a variety of prehistoric animals at their actual size. For each featured animal, the book includes some characteristics, how long ago it lived, and its length.

Boy, Were We Wrong About Dinosaurs!

By Kathleen Kudlinski, illustrated by S. D. Schindler
Dutton Children's Books, 2005
ISBN 978-0-525-46978-0
Grades 4–6

SYNOPSIS

This book examines what is known about dinosaur bones, behavior, and other characteristics. Kudlinski explains that as more evidence is discovered, our ideas about dinosaurs change.

Curricular Connections

According to the *National Science Education Standards*, young children should begin to understand that fossils are used by people today to learn about what Earth was like millions of years ago (NRC 1996). The American Museum of Natural History identifies four big ideas of paleontology on their "Paleontology: The Big Dig" website for kids. They are: (1) Fossils tell stories about Earth's history; (2) fossils can't tell us everything; (3) fossils are really rare; and (4) the fossil record is like a big jigsaw puzzle, with most of the pieces missing. It is not important that students memorize the names of and specific details about different dinosaurs or prehistoric animals; rather, the focus should be on these "big ideas." The activities here highlight a few of these big ideas of paleontology. In the lesson for K–2, students learn how scientists make inferences about prehistoric animals based on fossil evidence and by observing similar animals that are alive today. Students also learn what fossils can and can't tell us. In the lesson for grades 3–6, students learn how our ideas about dinosaurs change over time as more fossil evidence is discovered and how there is much more yet to be discovered! In this lesson, it is important to note that although all scientific ideas are tentative and subject to change, scientists do not change their ideas on a whim.

Grades K–2: Imagining Actual Size

Materials

- The Fossilizeded Remains of a Prehistoric Animal (p. 257)
- What Can Fossils Tell Us? T-chart (p. 258)

Engage

Show students the cover of *Prehistoric Actual Size* and ask what they think the book is about. Read page 3, which explains the premise of the book. Tell students that the animals are depicted "actual size," and the illustration may show just a part of an animal. They will have to use their imaginations to envision the rest. Read the book aloud and stop after reading about the terror bird on pages 26 and 27. Tell students that no person has actually seen any of the animals in this book. Ask, "If no one has ever seen these animals alive, how do we know how big they were?" Have them turn and talk to a partner and then share their ideas with the class. Explain that everything we know about the animals featured in this book is from fossils that people have found.

Explore

Give students a drawing of The Fossilized Remains of a Prehistoric Animal student page (p. 257). Ask students to observe the fossil picture and think about that animal's skin. What might it have looked like? What colors or patterns do you think it had? Have students use crayons to create the animals' outer form and skin. Then have students share and explain why they chose those particular colors and patterns. Ask, "Can anyone be sure of what color or patterns this animal had?"

Explain

Read the paragraph on page 28 of *Prehistoric Actual Size* that explains how scientists observe fossils and compare them to animals that are alive today to figure out what these animals looked like. After reading this section, project a T-chart that says "What Fossils Can Tell Us" on one side and "What Fossils Can't Tell Us" on the other side. Ask students what fossils can tell us about animals that lived long ago and record their responses in the first column of the T-chart. They should remember from the reading that fossils can give us a good idea of an animal's size, shape, how it moved, and what kinds of food it ate. Then ask them what fossils can't tell us and record their responses in the second column. They should explain that fossils can't tell us what color an animal was or whether it had spots, stripes, or other patterns. To make their best guesses about what colors and patterns the animals had, scientists compare them to similar animals that are alive today.

Mysteries of the Past

Elaborate

Revisit *Prehistoric Actual Size*. Choose a few of the animals that clearly resemble animals of today, such as the *Leptictidium* on pages 24 and 25, and ask students if they can think of any animals alive today that resemble it (e.g., *Leptictidium* has ears like a mouse, feet like a chicken, and body like a kangaroo). Reiterate that scientists used observations of similar animals that are alive today to make guesses about what colors and patterns these animals had.

Evaluate

Ask students why they think author Steve Jenkins called his book *Prehistoric Actual Size* and not *Prehistoric Actual Color*? Students should understand that fossils can give us a good idea of an animal's size but not the animal's color. Then ask students to explain what scientists can learn from fossil remains.

Grades 3–6: We Were Wrong About Dinosaurs!

Materials

- Jigsaw puzzle (one per each group of three to four students)
- Dinosaur Fossil Puzzle (p. 259)
- Boy, Were We Wrong About Dinosaurs! student page (p. 260)

Engage

Bring in several different jigsaw puzzles, one for each group of three to four students. Be sure to hide the lids. Give each group a few of the puzzle pieces. Ask students to discuss with each other and then share with the class what they think the whole picture looks like and why. Not only do students share their prior understanding by verbalizing and sharing with the teacher and the whole class, but this reflection also provides metacognition that will help other students build their own personal understanding. Next, give them a few more pieces of their puzzles and ask whether their ideas have

changed about what they think the whole picture looks like. Do this several times until they have most or all of the pieces. Ask, "How did your ideas change as you got each new piece of the puzzle?" Tell students that this is similar to how paleontologists work with fossils. The more pieces, or evidence, they find, the clearer their ideas are about what an animal looked like.

Explore

Cut apart the pieces of the Dinosaur Fossil Puzzle for each pair of students. Tell students that just like in the puzzle activity, paleontologists do not find complete dinosaur skeletons all at once. Instead, they must determine how the pieces go together, kind of like a puzzle. Most of the time, some of those pieces are missing. Give each pair of students one piece of the dinosaur puzzle. Have them discuss and draw what the dinosaur looked like based on that piece and then compare their drawings with other groups. Next, give them another piece and have them draw what they think the dinosaur looked like. Ask, "How have your ideas changed?" Continue until they have all the pieces. Tell students that this activity is similar to how paleontologists change their ideas about dinosaurs as more fossil evidence is discovered. But most of the time, paleontologists never find all the pieces!

Explain

Read *Boy, Were We Wrong About Dinosaurs!* aloud and ask students what they think is the author's most important message. Students should realize that the main idea of the book is that scientific ideas change as more evidence is discovered. And when you are talking about fossils, there's a lot more evidence out there yet to find. Ask them to list at least five examples from the book of how scientists' ideas about dinosaurs have changed on the Boy, Were We Wrong About Dinosaurs! student page (p. 260).

Elaborate

Give students time to further investigate dinosaurs and other prehistoric animals on the American Museum of Natural History website (see Internet Resources). Here they can learn how scientists create drawings based on fossils, watch interviews with paleontologists, act out a play about how animals become fossilized and how they end up in the museum, and vote in a "paleo poll." Students can also explore the museum's fossil exhibits at the AMNH Fossil Halls link.

Evaluate

As an assessment, check students' answers to the question at the bottom of the student page: "When you are reading books to find facts about dinosaurs, why is it important to check the copyright date?" (Students should demonstrate the understanding that as new fossils are found, scientists' ideas change. There are a lot of outdated books about dinosaurs!)

Reference

National Research Council (NRC). 1996. *National science education standards.* Washington, DC: National Academies Press.

Internet Resources

PaleontOlogy: The Big Dig
www.amnh.org/ology/index. php?channel=paleontology

The American Museum of Natural History Fossil Halls
www.amnh.org/exhibitions/permanent/fossilhalls

The Fossilized Remains of a Prehistoric Animal

Name: _____

What Can Fossils Tell Us?

What Fossils CAN Tell us	What Fossils CAN'T Tell us

Dinosaur Fossil Puzzle

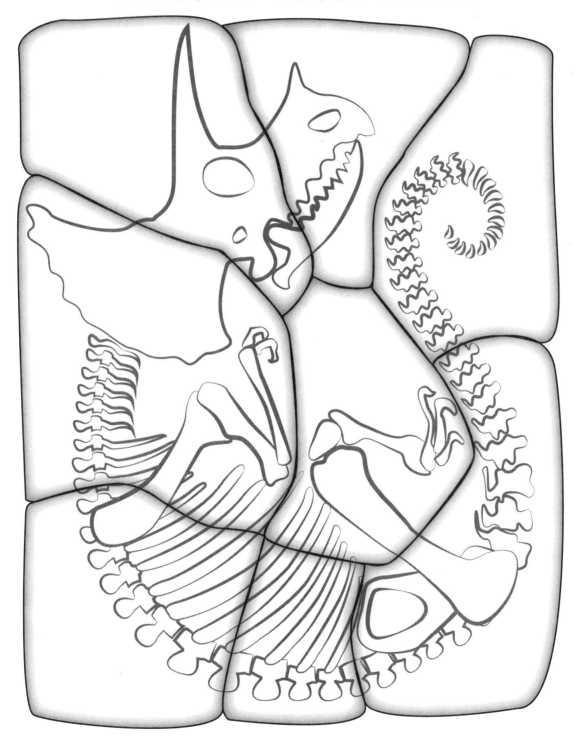

Name: _____

Boy Were We Wrong About Dinosaurs!

Directions: Use the T-chart below to list five ways, from the book *Boy, Were We Wrong About Dinosaurs!* that scientists have changed their ideas about dinosaurs as new evidence was discovered.

WE USED TO THINK	BUT NOW WE THINK
Example: Large bones discovered in ancient China belonged to dragons.	*Example:* Those large bones actually belonged to dinosaurs.
1	1
2	2
3	3
4	4
5	5

6 When you are reading books to find facts about dinosaurs, why is it important to check the copyright date? _____
_____
_____
_____

Chapter 46

Earthquakes!

By Emily Morgan and Karen Ansberry

Believe it or not, there is a 100% chance that an earthquake will happen somewhere in the world today. Although many of us think of Earth as solid ground, the crust is always moving. Most earthquakes are too small to feel, but we know they happen because of the sophisticated instruments we have to detect them. These activities use maps and models to help students better understand where and why earthquakes occur.

Trade Books

Earthquakes
By Ellen J. Prager, illustrated by Susan Greenstein
National Geographic, 2007
ISBN 978-1-4263-0090-5
Grades K–3

SYNOPSIS

From the Jump Into Science series from National Geographic, Earthquakes *uses simple text and illustrations to explain the reason for earthquakes, how often they happen, and where they are most likely to occur.*

Earthquakes
By Seymour Simon
HarperCollins, 2006
ISBN 978-0-06-087715-6
Grades 4–6

SYNOPSIS

This book from the Smithsonian series uses photographs, maps, diagrams, and text to explain the science of earthquakes.

Curricular Connections

The devastating earthquake of March 2011 in Japan has led elementary students to ask a lot of questions about earthquakes. The *National Science Education Standards* limit the topic of earthquakes in grades K–4 to the understanding that the surface of the Earth changes and that some of these changes are due to slow processes, like weathering and erosion, and others are due to rapid processes, such as volcanic eruptions and earthquakes (NRC 1996). In grades 5–8, the Standards introduce the concept of constantly moving lithospheric plates causing major geologic events, such as earthquakes. Both lessons here use interactive maps from the United States Geological Survey (USGS) website that show worldwide earthquake activity for the past week (see Internet Resources). From these maps, students can observe that hundreds of earthquakes take place every week and that there are certain places on Earth that experience earthquakes more often than others. In the K–3 lesson, students also build models to show that structures in these earthquake-prone areas should be attached to solid ground to prevent damage. In the lesson for grades 4–6, students determine earthquake activity in their area.

Grades K–3: Where Earthquakes Occur

Materials

- Photo from 1906 earthquake in Marin County, California (p. 265)
- Let's Learn About Earthquakes anticipation guide (p. 266)
- Metal pan
- Rock or brick
- Sand
- Clay

Engage

Show students the photograph of a fence offset by 8.5 ft. during the 1906 earthquake in Marin County, California. Ask students what they think might have happened to the fence. Tell students that the fence was moved during an earthquake. Ask students whether they have ever experienced an earthquake or heard about one, then have them share their experiences with a partner. Give each student a copy of the Let's Learn About Earthquakes anticipation guide (p. 266). In the "Before" column, have them record whether they think each statement is true or false. Tell students that they will have the chance to answer again at the end of the lesson. For younger students, you may want to answer the questions as a class.

Explore

Show students the USGS map titled "Latest Earthquakes in the World—Past 7 Days" (see Internet Resources). Read the title and then model how to use the map key. Show students that, according to the key, each box represents an earthquake that has happened in the past week; the larger the box, the stronger the earthquake. Ask students whether most of the boxes are small or large. Students will notice that most of the boxes are small. Tell students that most of these earthquakes were not strong enough to feel, but scientists have instruments called *seismographs* that can measure even the smallest earthquakes. Then show students the map of earthquake activity on page 17 of *Earthquakes* by Ellen J. Prager. Model how to use the map key to make meaning of the red and green dots on this map. Ask students whether they can name some places where earthquakes happen most often. Students should recognize from the map that earthquakes happen most often in or around the Pacific Ocean.

Explain

Show students the cover of *Earthquakes* by Ellen J. Prager. Ask students to signal when they hear an answer from the anticipation guide as you read the book aloud. Be sure to stop at pages 10 and 11, which feature illustrations of a fence being displaced by an earthquake and ask students whether the pictures remind them of something they have seen before. Students should make the connection to the photograph of the fence shown earlier. Show the photo again and compare it to the illustrations.

Elaborate

Reread pages 22 and 23 about how buildings and homes can be designed so they can withstand an earthquake without falling over or sinking into the ground. Tell students that to help understand this, we can use a model. Follow the procedure detailed on pages 30 and 31 of *Earthquakes* to create a model of a building built on sand. Using this model, students will observe that when a building is built on sand, it can sink down and tumble over during an earthquake. Next, clear away some sand and attach a small brick or rock securely to the bottom of the pan with clay. Tell students that this represents the building being attached to the solid ground. Shake the pan with a similar amount of force. The model should stay standing or not fall over as easily. Be sure to let students know that real buildings are not attached to solid ground with clay, but with structures like the ones featured on page 23 of the book.

Evaluate

Have students fill in the "After" column of the Earthquakes anticipation guide. Answers are as follows: (1) true, (2) false, (3) true, (4) true, (5) false, and (6) false. Discuss how their ideas have changed as they have learned about earthquakes. Model new learning by completing "I used to think …, but now I know …" statements. For example, "I used to think earthquakes did not happen very often, but now I know that earthquakes happen every day."

Grades 4–6: Why Earthquakes Happen

Materials

- Sticky notes
- Computers with internet connections or printed maps of Latest Earthquakes in the World—Past 7 Days
- Looking at an Earthquake Map O-W-L chart (p. 267)

Engage

Cover the title of the book *Earthquakes* by Seymour Simon with sticky notes, then show students the photograph on the cover and ask them what they think caused the destruction pictured. Students may say it was caused by a hurricane, tornado, or earthquake. Next, show them more photos from the book. Encourage them to revise their thinking as they see more photos. Afterward, tell students that all of the destruction shown in the photos was caused by earthquakes. Ask students whether they have ever experienced an earthquake or heard about one, and then allow them time to share their experiences with a partner.

Explore

Tell students that the USGS is a science organization that monitors earthquakes worldwide. Show students the USGS map titled "Latest Earthquakes in the World—Past 7 Days" (see Internet Resources). Model the basic map-reading skills necessary to make meaning from a map. Next, pair students on computers or give each pair a printed copy of the map. Then, give each student an O-W-L (Observations-Wonderings-Learnings) chart. Ask them to fill out the *O* column and share their observations of the map. Next, have students fill out the *W* (wonderings) column and share some of their wonderings with the class. Repeat this procedure with the USGS map of recent earthquakes in the United States. Next, have pairs of students choose

their most compelling wondering and write it on a sticky note. Post these questions in the classroom.

Explain

Tell students that the book *Earthquakes* by Seymour Simon might help answer some of their questions. As you read the book aloud, have students signal when they hear the answer to one of the wonderings posted in the room. After reading, have students fill in the *L* (learnings) column of the chart with these answers. Next, show students the Earthquakes 101 video, which includes computer-generated models of Earth's plate movements (see Internet Resources). Again, have students signal whether they hear any of the answers to their most compelling questions about earthquakes during the video and add these answers to the *L* column. Ask students whether any of their new learnings have led them to new wonderings. Tell them that often, the more you learn about something in science, the more questions you have. They can record any new wonderings in the *W* column of the chart.

Elaborate

Show students the picture of the seismograph reading on page 19 of *Earthquakes* and reread the two paragraphs on the opposite page about seismographs. Tell students that they can find the closest seismograph to their school by using a website called Rapid Earthquake Viewer (REV) (see Internet Resources). Have students click on "Station View" and enter the school zip code to find the closest seismograph. The readings for that day will show up immediately.

Evaluate

Have students add any additional learnings to the *L* column of their O-W-L chart and share them with the class. Then, using the USGS national and state maps, along with information they find at the REV site, have them determine whether your area is at high risk or low risk for earthquakes. Students can communicate their answer in writing or orally, but must include at least five pieces of evidence to support their answer.

Reference

National Research Council (NRC). 1996. *National science education standards*. Washington, DC: National Academies Press.

Internet Resources

National Geographic Earthquakes 101 Video
http://video.nationalgeographic.com/video/player/ environment/environment-natural-disasters/ earthquakes/earthquake-101.html

Photograph of Fence Offset by 1906 Earthquake
http://earthquake.usgs.gov/regional/ nca/1906/18april/images/fenceoffset_big.html

Step-by-Step Instructions for Using the Rapid Earthquake Viewer by IRIS
www.iris.edu/hq/files/publications/brochures_ onepagers/doc/REV_Insert.pdf

The Rapid Earthquake Viewer by IRIS (Incorporated Research Institutions for Seismology)
http://rev.seis.sc.edu

USGS Latest Earthquakes in the World—Past 7 Days
http://earthquake.usgs.gov/earthquakes/ recenteqsww

Science 101: How Are Skyscrapers Designed?
www.nsta.org/publications/article.aspx?id=Z349U Ri8cV6FALDuNq!plus!8OLly!plus!tmivolv!plus!TlsX Z4KYcs=

Earthquakes!

Photo of a fence offset by 8. ft. during the 1906 earthquake in Marin County, California.

From: *http://earthquake.usgs.gov/regional/nca/1906/18april/images/fenceoffset_big.html*

Name: _____

Let's Learn About Earthquakes

Anticipation Guide

Before **True or False**		*After* **True or False**
_____	1. Earthquakes happen every day.	_____
_____	2. Rocks are too hard to ever break.	_____
_____	3. Earthquakes last a few seconds or minutes.	_____
_____	4. Some places on Earth have more earthquakes than other places.	_____
_____	5. Buildings should be built on sand to keep them from falling during an earthquake.	_____
_____	6. We can stop earthquakes from happening.	_____

Name: _____

Looking at an Earthquake Map

O	W	L
What do you **OBSERVE** about earthquakes?	What do you **WONDER** about earthquakes?	What did you **LEARN** about earthquakes?

Chapter 47

Delving Into Disasters

By Christine Anne Royce

T he topic of natural disasters provides many avenues for students to delve into existing data and conduct research. The activities described here involve students in uncovering information about previous natural disasters or events to construct explanations and use the science process skills of observation, prediction, and drawing conclusions. Students are then better able to consider what type of information they need to research a topic.

⊃ Trade Books ⊂

The Blizzard
By Betty Ren Wright, illustrated by Ronald Himler
Holiday House, 2005
ISBN 978-0-8234-1981-4
Grades K–3

SYNOPSIS
This book allows children to imagine what it was like in the early 20th century when a blizzard hit. Billy was waiting for his cousins to visit on his birthday, so most of the events transpire at home during the blizzard; however, the teacher can easily elicit students' responses related to the amount of snow falling and what they understand about blizzards by connecting it to students' lives.

Hurricanes!
By Gail Gibbons
Holiday House, 2010
ISBN 978-0-8234-2297-5
Grades 2–5

SYNOPSIS
Gail Gibbons uses clear and simple language to introduce the topic of hurricanes. She describes the different categories of hurricanes, how they form, their structure, and the aftermath of a hurricane hitting an area.

Curricular Connections

Using data in inquiry investigations is important to helping students engage in science. Data can be collected while doing an activity and then interpreted, or existing data can be examined for patterns. These activities use existing data about natural disasters to help students make observations about the data and draw conclusions. The *National Science Education Standards* emphasize that students should be able to "use data to construct a reasonable explanation" and that "even at the earliest grade levels, students should learn what constitutes evidence and judge the merits or strength of the data and information that will be used to make explanations" (NRC 1996, p. 122).

These activities have students looking at data about snowfall, rainfall, and hurricanes to consider where these events occur and what types of conclusions we can draw about natural disasters. For the K–3 activity, students are asked to consider snowfall (or rainfall) in their local area over time. Even though this data is preexisting, it will require them to organize it in a meaningful way and then be able to draw some general conclusions about snowfall or rainfall in their area.

Students in grades 4–6 examine the paths some famous hurricanes have taken. They can then make observations about this data—one factor associated with how hurricane researchers predict the paths of future storms.

Grades K–3: Snow/Rain Trackers

Purpose

Students will investigate the average annual snowfall (or rainfall) in a given area.

Materials

*   Internet access
*   Graph paper
*   Snowfall/Rainfall Data Table (p. 273)

Procedure

1. Before doing this activity, research the amount of snow or rain your particular area has received each year for the past 10 years (try your local weather station). Complete the snowfall (or rainfall) data table (p. 273).

2. Read *The Blizzard* to the class and discuss the story. (If you live in an area that does not get snow, adapt this lesson to discuss heavy rainstorms.) The discussion may initially focus on how Billy felt when his cousins couldn't get through the snow to visit for his birthday, but the teacher should eventually move the discussion toward asking the ways in which weather has affected students' lives. What do they do on a stormy day?

3. As students discuss the different snowstorms (or rainstorms) they have experienced, ask them how much snow they think their area has gotten each year since they were born. Students will have descriptive words such as "lots" and "piles" but not necessarily quantitative terms. Ask the students to consider how researchers keep track of snowfall amounts over time.

4. Present the information you compiled to the students and ask them to discuss what kind of conclusions they can draw from the data. Ask, "Does any of the data stand out? Do you notice any patterns?"

5. Have students think about the data in a different way, and ask them to consider how they could make a graph about the data and begin to examine the information. Young students can develop either a pictograph or bar graph representing inches of snow per year. Ask questions associated with the graph such as, "What year had the most snow?" "What year had the least snow?" and "Does the graph help to show if there was a blizzard in a given year?" (The graph actually doesn't demonstrate that, which is why we need

Delving Into Disasters

to keep different types of data—such as wind speed— about individual events as well.)

6. Students can even make predictions about how much snow might fall in the current year. If teachers have them do this, students should be asked to support their predictions with explanations as to why they stated that.

Allowing students to examine data and draw some conclusions associated with the data engages them in part of the inquiry process associated with research. To extend this activity, have students keep track of snowfall or rainfall amounts each day over a period of time and graph those results that represent actual daily data rather than aggregated data. An adaptation of this activity would be to gather the information about snowfall or rainfall for a single year using monthly averages rather than annual averages.

Grades 4–6: Hurricane Happenings

Purpose

Students will research and track the movement of a hurricane to determine what path it takes.

Materials

- Hurricane tracking chart (See Internet Resources.)
- Data from a hurricane (pp. 274–277)
- Colored pencils or markers

Procedure

1. Begin by asking students if they have ever heard of a hurricane and what it might be like. Prompting questions like, "Can you describe what a hurricane is?" or, "Where do most hurricanes occur?" will start the discussion with the students.

2. Use Gail Gibbons's book *Hurricanes!* as a read-aloud for the class, making sure to stop and gen-erate a list of the keywords on the board, such as *tropical depression* and *storm surge*. Although defining vocabulary is not the main goal of this activity, these terms will become important as the students begin to research their own hurricane and plot it on a tracking chart.

3. Using a transparency of a hurricane tracking chart (see Internet Resources), ask the students to make predictions about where hurricanes start and what path they may take.

4. Next, review map skills. Using the data from a single hurricane (see Internet Resources), model how to locate latitude and longitude on the map and mark it with a day and time. Each day/time entry represents a data point from the storm. Once the points have been plotted, demonstrate how to draw or follow the path of the storm by connecting the data points. An example piece of data may look like the following: On September 1, Hurricane X was located at 24N latitude and 70W longitude at 4:50 p.m., and the classification of the storm was tropical storm, tropical depression, or hurricane. Other information that accompanies hurricane data includes wind speed, storm surge, and speed of movement. Although this information does help researchers predict the path of the storm, it will not be used for this simplified version of predicting the path.

5. Explain to the students that the National Hurricane Center keeps data about each hurricane that has already occurred and that they will be examining the data and "tracking" their own storm. Pairs of students can track the same storm or different storms. Provide each student or pair of students with one of the data sheets for either Hurricane Felix or Hurricane Stan (pp. 274–277). Information for two different storms has been distilled for use with students at this age level. Depending on the attention level and age of your students, you

may want to either limit the pieces of data to plot or do this as a large class activity.

6. Once students have successfully plotted the points associated with their hurricane, ask them to draw some conclusions about the storm. Where did it originate? What path did it take? Did it go over land or stay over water? Did the path change direction at any point? If so, what might have caused the change in direction?

7. Allow the students to consider the following question as you return to the book: Why is it important for meteorologists to track and predict hurricanes?

Although no one wants a disaster to happen, there is a definite source of information available for students to use when considering how to use data in science and collect data for research purposes.

Reference

National Research Council (NRC). 1996. *National science education standards.* Washington, DC: National Academies Press.

Internet Resources

Atlantic Basin Hurricane Tracking Chart
  www.nhc.noaa.gov/pdf/tracking_chart_atlantic.pdf
National Hurricane Center Data Archive
  www.nhc.noaa.gov/pastall.html

Snowfall/Rainfall Data Table

YEAR	INCHES OF SNOW	ANY BLIZZARDS?

Hurricane Felix

August 31 – September 5, 2007

Best track for Hurricane Felix, August 31–September 5, 2007.

DATE/TIME (UTC)	LATITUDE (°N)	LONGITUDE (°W)	PRESSURE (MB)	WIND SPEED (KT)	STAGE
31 / 1200	11.5	56.6	1009	25	tropical depression
01 / 0000	12.1	59.4	1007	35	tropical storm
01 / 1200	12.2	62.8	1001	50	tropical storm
02 / 0000	12.6	66.1	992	65	hurricane
02 / 1200	13.0	69.4	980	90	hurricane
03 / 0000	13.8	73.0	935	150	hurricane
03 / 1200	14.2	76.9	937	140	hurricane
04 / 0000	14.4	80.4	950	115	hurricane
04 / 1200	14.3	83.2	934	140	hurricane
05 / 0000	14.6	85.4	982	50	tropical storm
05 / 1200	15.5	87.3	1006	20	remnant low
06 / 0000	16.5	89.5	1007	20	remnant low
06 / 1200	17.2	92.4	1007	20	remnant low
07 / 0000					dissipated

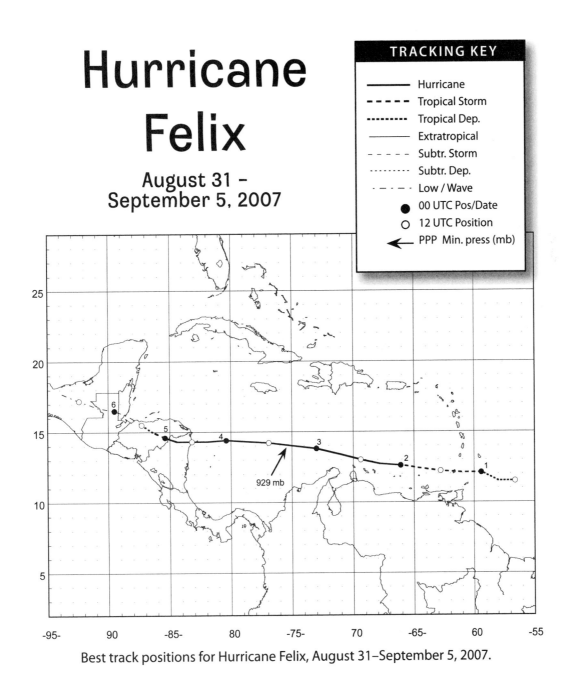

Hurricane Felix

August 31 – September 5, 2007

TRACKING KEY

— Hurricane
- - - Tropical Storm
···· Tropical Dep.
— Extratropical
– – – Subtr. Storm
······ Subtr. Dep.
–·–· Low / Wave
● 00 UTC Pos/Date
○ 12 UTC Position
← PPP Min. press (mb)

929 mb

Best track positions for Hurricane Felix, August 31–September 5, 2007.

Hurricane Stan

October 1–5, 2005

Information distilled from
http://www.nhc.noaa.gov/pdf/TCR-AL202005_Stan.pdf

DATE/TIME (UTC)	LATITUDE (°N)	LONGITUDE (°W)	STAGE
October 01/ 1200	18.9	85.6	tropical depression
October 02/ 0000	19.3	86.7	tropical depression
October 02/ 1200	19.8	87.9	tropical depression
October 03/ 0000	20.5	89.8	tropical depression
October 03/ 1200	20.3	91.7	tropical depression
October 04/ 0000	19.8	93.2	tropical depression
October 04/ 1200	18.6	94.9	tropical depression
October 05/ 0000	17.2	96.4	tropical depression
October 05/ 1200			dissipated

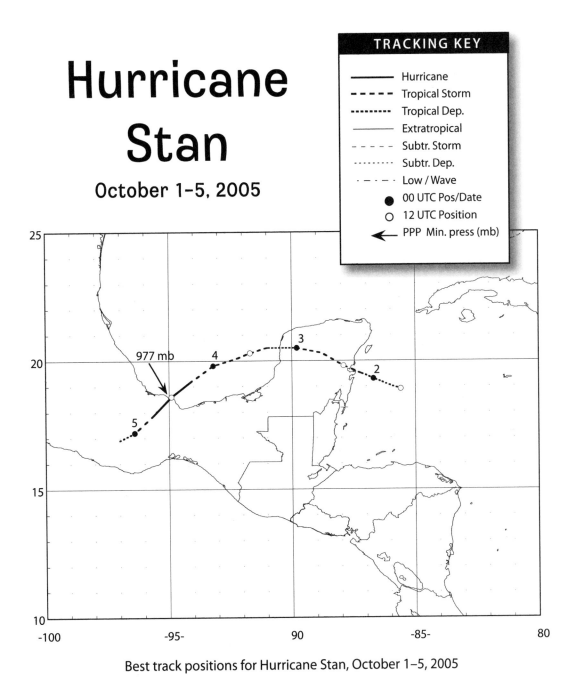

Hurricane Stan

October 1–5, 2005

TRACKING KEY

——	Hurricane
– – –	Tropical Storm
········	Tropical Dep.
——	Extratropical
– – –	Subtr. Storm
·········	Subtr. Dep.
–·–·	Low / Wave
●	00 UTC Pos/Date
○	12 UTC Position
←	PPP Min. press (mb)

977 mb

Best track positions for Hurricane Stan, October 1–5, 2005

Chapter 48

Cloud Watchers

By Emily Morgan, Karen Ansberry, and Colleen Phillips-Birdsong

Weather is a topic in science that is applicable to our lives on an everyday basis. The weather often determines what we wear, where we go, and what we do. The activities here focus on clouds and the part they play in determining our weather. In the K–3 lesson, students learn about different cloud types and sculpt each type out of shaving cream. In the lesson for grades 4–6, students learn about Luke Howard—the man responsible for naming clouds—and then investigate clouds by collecting real data for NASA's S'COOL project (see Internet Resource).

Trade Books

Fluffy, Flat, and Wet: A Book About Clouds

By Dana Meachen Rau, illustrated by Denise Shea
Picture Window Books, 2006
ISBN 978-1-4048-1729-6
Grades K–3

SYNOPSIS

Simple text and colorful illustrations help students understand what clouds are made of, how clouds form, and the differences among cloud types. The author includes fun facts and a cloud journal activity.

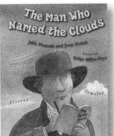

The Man Who Named the Clouds

By Julie Hannah and Joan Holub, illustrated by Paige Billin-Frye
Albert Whitman & Company, 2006
ISBN 978-0-8075-4974-2
Grades 4–6

SYNOPSIS

This book details the life and work of Luke Howard, who loved clouds and weather so much that he created the first practical scientific system for naming them.

Curricular Connections

The *National Science Education Standards* (*NSES*) suggest that students in grades K–4 observe the sky regularly, write and draw descriptions, and identify patterns (NRC 1996). Young students should understand that objects in the sky have properties and movements that can be observed and described and that weather changes from day to day and from season to season. The most important part of this lesson is the general understanding that different types of clouds bring different types of weather. In this K–3 lesson, students spend time observing clouds and recording their wonderings, read about the different types of clouds, and make models of the three main cloud types. Students are then introduced to the type of weather associated with each cloud. In the lesson for grades 4–6, students not only learn about different types of clouds and the weather associated with them, but they also learn about an important person in weather history, Luke Howard. Through reading about Luke Howard, students understand the need for a cloud classification system and the process Howard used in developing this system. The *NSES* suggest that by studying individuals that have contributed to the traditions of science, students can gain further understanding of scientific inquiry, the nature of science, and the relationship between science and society. Finally, students apply what they have learned about clouds and the nature of science to become part of a NASA project to study how clouds may affect Earth's climate.

Grades K–3: Cloud Sculpting

Materials

- Classroom-size O-W-L chart
- Shaving cream (not gel)
- Thermometer
- Cloud Watchers overhead (p. 283)
- My Weather Forecast student page (p. 284)

Engage/Explore

Take students outside and give them a few minutes to observe the clouds silently. Then allow them to talk to a partner about what they observed. Return to the classroom and give each pair of students a photo set of various cloud types (see Internet Resources). Next, gather students around a cloud-shaped O-W-L (Observations, Wonderings, Learnings) chart. Ask them to share their observations about clouds. Write their observations in the O section of the chart. Then explain that observations often lead to questions or "wonderings." Write some wonderings—such as, "What are clouds made of?"—in the W column of the chart. Ask students what questions they have about clouds. Record some of their questions in the W section of the chart.

Explain

Tell students that you have a nonfiction book to help them answer some of their questions about clouds. As you read *Fluffy, Flat, and Wet: A Book About Clouds,* have them signal when they hear the answer to one of the questions from the *W* section of the O-W-L chart. After reading, add to the *L* column of the chart. This chart can be posted in the room and added to.

Elaborate

Tell students that they are going to sculpt three of the cloud shapes that they learned about in the book. Squirt a mound of shaving cream (not gel) directly onto the desk in front of each student (check to be sure students do not have allergies or sensitivities in advance). Students should wear goggles for safety. First, have the students sculpt their shaving cream into fluffy, rounded clouds. Underneath their cloud sculptures, have them spread a thin layer of shaving cream and write the word *cumulus*. Explain that cumulus clouds bring different kinds of weather depending on where they are in the sky: low (bright, sunny), middle (thunderstorms), high (getting colder). Next, have students sculpt flat, blanketlike clouds with the word *stratus* written in a thin layer of shaving cream. Explain that stratus clouds bring different kinds of weather depending on where they are in the sky: low (a little rain), middle (rain), high (rain or snow). Finally, have students sculpt wispy, featherlike clouds and write the word *cirrus* in shaving cream. Explain that cirrus clouds occur high in the sky on sunny, dry days. Explain that there are other types of clouds as well.

Evaluate

Now students are going to use what they have learned about clouds to be meteorologists, or scientists who study the weather. Observe clouds again, this time three different times in one day, and record the air temperature each time. Project an image of the Cloud Watchers page (p. 283) and record the students' observations. You may want to watch NASA's Cloud Types Tutorial (see Internet Resources) to assist students in determining cloud cover. Then, analyze the data as a class. What do they think the weather will be and why? Tell students that clouds are just one thing that meteorologists look at when predicting weather (they also look at air pressure and wind speed). Next, pass out the My Weather Forecast student page (p. 284) and have each student make a weather forecast based on their cloud observations and temperature measurements. Have each student make a picture in the box showing what they think the weather will be and draw themselves dressed for the weather.

Grades 4–6: Cloud Watching

Materials

- Cloud Watchers Journals (pp. 285–286)
- Photos of various cloud types
- S'COOL Report Form (p. 287)

Engage

Give each student a copy of the Cloud Watchers Journal (pp. 285–286) and have them keep observations of clouds for at least two weeks. Students should record the date, time, description of the clouds they observe, and drawings of the clouds. Have students share their journals in pairs and then come up with a few questions to share. Students can write these "Cloud Wonderings" on sticky notes and post them on a cloud-shaped chart in the classroom.

Explore

Ask students whether they have ever heard names for different kinds of clouds. Have them turn and talk with a partner about any cloud names they may have heard. Next, provide students with photo sets of various cloud types (see Internet Resources). Ask them to sort the photos into groups and come up with names for them based on their observations. Have students share the names they have come up with. (Be sure to number each photo in advance so that students in different groups can know which photo is being described by another group.)

Explain

Tell students that you have a story to share about a boy who loved to look at the clouds. Show them the cover of *The Man Who Named the Clouds*. Point out that the man on the cover is writing in a weather journal, just as they have done. Tell them to listen for the names of three main cloud types and what those names mean. Read the book aloud, skipping the Weather Journal insets to read at another time. After reading, have students recall the three main types of clouds: cirrus (which means *curl of hair*), cumulus (which means *heap*), and stratus (which means *layer*). Ask students how these names compare to the names they came up with. Next, have them classify their cloud pictures into those three categories. You may also want to show them NASA's cloud types tutorial (see Internet Resources).

Elaborate/Evaluate

To give your students a real-life science connection to what they have learned about clouds, you can register your class in NASA's S'COOL project. Students worldwide are making *ground truth measurements* for the Clouds and the Earth's Radiant

Cloud Watchers

Energy System (CERES) experiment. Ground truth measurements are land-based observations to compare with satellite data for the purpose of improving the satellite results. Scientists are using the satellite data to study how clouds may affect Earth's climate. Participants are asked to make some basic weather observations and to record the type and features of clouds in the sky at the time that one of the satellites carrying CERES instruments passes over their location. The S'COOL website has a variety of tutorials and other materials to help prepare students to make these observations and enter them into NASA's database. Teachers can use the S'COOL Report Form (p. 287) to evaluate student learning.

Reference

National Research Council (NRC). 1996. *National science education standards.* Washington, DC: National Academies Press.

Internet Resources

Cloud Types Tutorial
http://science-edu.larc.nasa.gov/SCOOL/tutorial/clouds/cloudtypes.html

NASA's S'COOL Project
http://scool.larc.nasa.gov

Photos of Various Cloud Types
http://science-edu.larc.nasa.gov/SCOOL/cldchart.html

Cloud Watchers

Weather Forecasting Chart					
Cumulus		**Stratus**		**Cirrus**	
If the cloud is:	*The weather could be:*	*If the cloud is:*	*The weather could be:*	*If the cloud is:*	*The weather could be:*
Low	Sunny	Low	A little rain	High	Sunny, dry
Medium	Thunderstorms	Medium	Rain		
High	Getting colder	High	Rain or snow		

Our Cloud Observations for _____ (date)

Observation #1   Time: _____

Drawing	Shape	Level	Color	Temp.	Rain?
	Cumulus Stratus Cirrus	Low Medium High	White Light Gray Dark Gray		Yes No

Observation #2   Time: _____

Drawing	Shape	Level	Color	Temp.	Rain?
	Cumulus Stratus Cirrus	Low Medium High	White Light Gray Dark Gray		Yes No

Observation #3   Time: _____

Drawing	Shape	Level	Color	Temp.	Rain?
	Cumulus Stratus Cirrus	Low Medium High	White Light Gray Dark Gray		Yes No

My Weather Forecast

for _____ by _____
        (date)   *(meteorologist's name)*

I think the weather will be

_____

because_____

_____.

Draw what you think the weather will look like. Draw yourself dressed for the weather in the picture!

Cloud Watchers Journal

by: _____
meteorologist's name

Observation #1

Date: _____ Time: _____
Observations: _____
_____
Sketch:

Observation #2

Date: _____ Time: _____
Observations: _____
_____
Sketch:

Observation #3

Date: _____ Time: _____
Observations: _____
_____
Sketch:

Observation #4

Date: _____ Time: _____
Observations: _____
_____
Sketch:

Observation #5

Date: _____ Time: _____
Observations: _____
_____
Sketch:

Observation #6

Date: _____ Time: _____
Observations: _____
_____
Sketch:

Cloud Watchers Journal

by: _____

meteorologist's name

Observation #7
Date: _____ Time: _____
Observations: _____
_____
Sketch:

Observation #8
Date: _____ Time: _____
Observations: _____
_____
Sketch:

Observation #9
Date: _____ Time: _____
Observations: _____
_____
Sketch:

Observation #10
Date: _____ Time: _____
Observations: _____
_____
Sketch:

Observation #11
Date: _____ Time: _____
Observations: _____
_____
Sketch:

Observation #12
Date: _____ Time: _____
Observations: _____
_____
Sketch:

S'COOL Report Form

Login ID: _____ City: _____

Date (ex. 2001 09 20): Year _____ Month _____ Day _____ Satellites _____

Local Time *(24 hour Clock ex 14 26)*: Hour _____ Minute _____ Universal Time: Hour _____ Minute _____

Cloud Observations: *(Select the most prevalent cloud type at each level where clouds exist. Cloud Cover and Visual Opacity must be determined for each level observed. Use the comment section for further descriptions.)*

☐ Clear Sky - no Clouds observed *(skip to the "Surface Cover" section)*
☐ Clouds Present - *(continue to level(s) observed - don't forget to count contrails if present)*

High Level

Cloud Type:	Cloud Cover:		Visual Opacity:
☐ Cirrus	☐ Clear	(0-5%)	☐ Opaque
☐ Cirrocumulus	☐ Partly Cloudy	(5% - 50%)	☐ Translucent
☐ Cirrostratus	☐ Mostly Cloudy	(50% - 95%)	☐ Transparent
	☐ Overcast	(95% - 100%)	

Mid Level

Cloud Type:	Cloud Cover:		Visual Opacity:
☐ Altrostratus	☐ Clear	(0-5%)	☐ Opaque
☐ Altocumulus	☐ Partly Cloudy	(5% - 50%)	☐ Translucent
	☐ Mostly Cloudy	(50% - 95%)	☐ Transparent
	☐ Overcast	(95% - 100%)	

Low Level

Cloud Type:		Cloud Cover:		Visual Opacity:
☐ Fog	☐ Stratus	☐ Clear	(0-5%)	☐ Opaque
☐ Nimbostratus	☐ Cumulus	☐ Partly Cloudy	(5% - 50%)	☐ Translucent
☐ Cumulonimbus	☐ Stratocumulus	☐ Mostly Cloudy	(50% - 95%)	☐ Transparent
		☐ Overcast	(95% - 100%)	

Ground Observations

Surface Cover: *(Mandatory)*

Yes	No	
☐	☐	Snow/Ice
☐	☐	Standing Water
☐	☐	Muddy
☐	☐	Dry Ground
☐	☐	Leaves on Trees
☐	☐	Raining/Snowing

Surface Measurments: *(Optional–you may submit any or all)*

Temperatures: Barometric Pressure: (Select one)

_____ Celsius or   _____ Celsius or   _____ hPa   _____ psi
_____ Fahrenheit   _____ Fahrenheit   _____ mb   _____ inches Hg
                                          _____ atm   _____ torr (mm Hg)

Relative Humidity: _____ %

Chapter 49
Weather Watchers

By Christine Anne Royce

Students probably have heard weather-based sayings, such as "March comes in like a lion and goes out like a lamb" or "April showers bring May flowers." Throughout the ages, people have developed these and other sayings to try to predict what the weather holds for them, their locations, and their lifestyles. Even so, few people truly understand what causes the weather. Making weather observations with students provides opportunities to introduce meteorology while helping students develop their observation and data-collection skills.

Trade Books

The Cloud Book
By Tomie dePaola
Holiday House, 1984
ISBN: 978-0-8234-0531-2
Grades K–3

SYNOPSIS
Although more than 25 years have passed since its original publication, The Cloud Book *is a classic for helping students learn about the weather. This illustrated book provides students with information about different types of clouds, the weather associated with each cloud, and some of the sayings people use in the description of clouds. This book is easy to read and provides a sufficient amount of background information on clouds.*

Weather Forecasting
By Gail Gibbons
Aladdin, 1993
ISBN: 978-0-689-71683-6
Grades 2–5

SYNOPSIS
Weather Forecasting *introduces the reader to meteorologists and the various instruments used to gather information to predict future weather. In narrative format, the book explains what a meteorologist does while also including many illustrations and graphics with facts about the equipment used in weather forecasting.*

Curriuclar Connections

Weather is a simple term for the processes that govern the Earth's atmosphere—meteorology. Weather occurs as a result of the Sun heating the ground and the ground heating the atmosphere. As the air near the ground is warmed, it rises. Rising air cools and if the air is moist enough, clouds may form. Variations in temperature, along with the rotation of the Earth, help form the wind systems and patterns that develop throughout the world and bring each region its weather.

One of the most easily recognizable signs of weather are clouds. There are three main types of clouds—*cirrus, stratus*, and *cumulus*. Various prefixes and suffixes, such as *strato, alto,* and *nimbus*, as well as the act of combining cloud types, bring the total types of clouds up to 10.

Clouds form when water vapor in the air cools to a point that it begins to condense around tiny particles called *condensation nuclei*. Air can cool through the process of conduction by passing over ground or water that is cooler, or it can rise and cool as it uses up energy. The amount of water vapor in the air, its altitude, and the strength of rising air currents determine the type of cloud(s) formed.

The types of clouds present and how much of the sky they cover are only some of the clues meteorologists use to forecast the weather. Other pieces of information that are collected include wind speed, wind direction, air temperature, humidity, barometric pressure, and precipitation. All of these components together help form the complex system we call *weather*.

The individual characteristics of a location coupled with atmospheric conditions produces the weather for each individual region. As students practice making observations and noting the characteristics of the weather, they will become more adept at forecasting the weather in their hometown. Be sure to let them know that forecasting is just another type of prediction—and that not every prediction will be correct or accurate!

Grades K–3: Crazy About Clouds!

Purpose

Students will have the opportunity to observe and identify clouds.

Materials

- Chart paper
- Cloud Observation Forms (p. 295)
- Craft materials such as yarn, glue, cotton balls, and markers

Procedure

1. Young students have terrific observation skills—especially when it comes to nature. This activity allows students to use these skills while increasing their science knowledge about clouds and the weather. Begin by reading *The Cloud Book* to the class and asking specific questions to draw on students' prior knowledge and observations: "Have you ever walked outside and looked up?" "What do you see in the sky?" and "Do all clouds look the same?" Throughout the reading, such questions as "What are the three main types of clouds?" and "How would you describe each type of cloud?" will help focus the student's learning of material presented in the text.

2. At the conclusion of the book, create a large chart with the help of the students. Identify names of cloud types, what they look like, and the type of weather the cloud indicates. This will reinforce the content material the students heard in the story and will serve as an instructional aid for the activity. Have the students construct a "cloud book" of their own in which they will record their observations. Tie together several copies of the Cloud Observation Form with yarn or string, and add a cover decorated by each student using cotton balls or markers.

Weather Watchers

3. Now that the stage has been set, provide an opportunity for students to visit a quiet place with an unobstructed view of the sky each day to observe the clouds. Ask students to draw the types of clouds they see and try to name them, using the classroom chart to refresh the students' memories. Each day's observations should be drawn on a single page of the cloud journal. As students are working, the teacher can easily assess if the student understands the type of clouds and, if necessary, probe deeper into the student's knowledge by asking individual questions.

4. Extensions might include having students record the weather they observe and perhaps a prediction for tomorrow's weather. This activity not only helps meet the *National Science Education Standards* of observing and describing objects in the sky and describing weather by "measurable quantities, such as temperature, wind direction and speed, and precipitation" (NRC 1996), it also provides the opportunity to integrate language arts activities through writing and giving oral reports on their observations. A follow-up story could be *Cloud Dance* by Thomas Locker (2003), which includes colorful paintings of different cloud types accompanied by poetic lyrics.

Grades 4–6:
Weather Forecasting

Purpose

Using simple instruments, students will gather information and make weather forecasts.

Materials

- Rain gauge
- Thermometer
- Barometer
- Wind Vane
- Weather Reporting Form (p. 296)

Procedure

1. For older students, setting up a classroom weather station is a great way for students to practice collecting data on the weather and then making predictions based on their data. You will need to either obtain a classroom weather station with instruments provided or collect the materials to build simple weather instruments. Depending on your students, you may wish to have the students build the instruments themselves. The book *Weather Forecasting* is a great place to start to learn more about the tools used in monitoring the weather.

   A thermometer should be placed outdoors (not in direct sunlight) for the most accurate reading or be available for students to take outdoors to obtain a reading. Barometers measure atmospheric pressure—an important piece of information in forecasting future weather.

   See sidebars for directions for building a classroom barometer and a wind vane (pp. 292–294). An inexpensive rain gauge can be made using any straight-sided cylinder and ruler. The rain gauge should be placed in an unsheltered location on level ground. Students should record the amount of precipitation at the same time each day and then empty the rain gauge.

2. Once the classroom weather station is established, students can collect data about the weather on a daily basis. A Weather Reporting Form (p. 296), can assist in keeping track of the data. This activity can be conducted throughout the month by assigning pairs of students an individual day on which to collect the data and post it in a classroom area dedicated to weather. As part of their daily science lesson, students could predict tomorrow's weather based on the data.

3. Weather forecasters use not only their science knowledge, but also a variety of subjects to prepare their daily reports. Predictions or forecasts could be written in the form of a newspaper report or videotaped as in a television report,

allowing teachers to integrate the language arts. Some schools may even allow students to broadcast the weather during the morning announcements as a service to the school.

4. Math could be integrated by having students graph daily precipitation amounts or the number of sunny, partly sunny, cloudy, and rainy days in a month. Tracking the changes in the weather lends itself to various activities within the curriculum.

References

Locker, T. 2003. *Cloud dance.* New York: Voyager.

National Research Council (NRC). 1996. *National science education standards.* Washington, DC: National Academies Press.

Internet Resources

National Oceanic and Atmospheric Administration *www.noaa.gov*

National Oceanic and Atmospheric Administration Education Resources *www.education.noaa.gov*

How the Weatherworks *www.weatherworks.com/monthly/activities/ activity_index.html*

Making and Using a Barometer

By Marvin N. Tolman and Garry R. Hardy

The difficulty in using a homemade barometer is that it responds to changes in air temperature as well as atmospheric pressure, and we assume the change is due to ups and downs of air pressure. This can really mess up the data on a weather chart and make it difficult for students to establish atmospheric pressure trends and their relationship to weather patterns.

First, to make a homemade barometer that will enable you and your students to report changes in atmospheric pressure in the classroom, you will need the following:

- Wide-mouth glass jar, such as a mayonnaise jar, or other container with rigid, nonflexible sides
- Large, round balloon (at least 22 cm in diameter when inflated)
- Strong rubber band or string (to secure the balloon to the jar)
- Commercial barometer
- Drinking straw
- Index card or sheet of paper
- Thermometer
- Scissors

To make the barometer, cut the narrow neck off the balloon and stretch the remaining part tightly over the mouth of an open-top glass jar. The balloon now forms a diaphragm over the top of the jar. Secure the balloon to the jar with a rubber band. To make the "pointer," cut one end of the straw into a point and glue the other end of the straw to the center of the diaphragm in a horizontal position. Your barometer is now complete.

To use the barometer, attach an index card or a piece of paper to the wall or a stand that can be placed near the barometer's pointer. Make a mark on the card to show the pointer's current position. Consult your commercial barometer or call the local weather station to determine the current barometric pressure (atmospheric pressure), and record this number beside the mark on your index card.

Every day for one week (or more) mark the position of the pointer and record the current barometric pressure beside it (see Figure 49.1). You will soon be able to put away the commercial barometer and use the homemade barometer to report changes in barometric pressure. Close supervision is required for safety reasons; the glass jar could cause injury if broken.

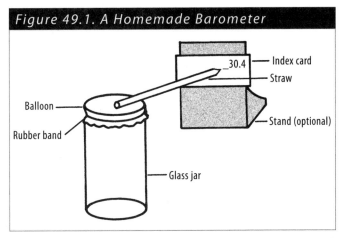

Figure 49.1. A Homemade Barometer

The numbers representing barometric pressure are less important than the direction of movement of the pointer. Generally, decreasing barometric pressures accompany storm fronts, while increasing pressures are associated with fair weather conditions.

When the diaphragm is first stretched over the jar, the air pressure in the jar is the same as that of the atmosphere in the room. As atmospheric pressure increases or decreases, it changes the amount of pressure exerted on the surface of the diaphragm, causing the straw to move up or down. As atmospheric pressure increases, the diaphragm is pressed down, causing the pointer to rise, thus signaling an increase in atmospheric pressure. Conversely, a drop in the position of the pointer indicates decreasing atmospheric pressure.

The difficulty in using a homemade barometer occurs when the temperature of the air trapped inside the jar gets warmer. As the trapped air gets warmer, it expands, causing the diaphragm to rise, the pointer to drop, and the barometer to give a false reading of a drop in atmospheric pressure. As the room air cools, the temperature of the trapped air under the diaphragm also cools, causing the trapped air to condense and decreasing the pressure inside the barometer. The diaphragm drops and the pointer rises, leading one to think the barometric pressure is going up when in fact the barometer is really responding to change in air temperature. As air temperature changes, the "barometer" behaves as a "thermometer."

To avoid false readings, keep a thermometer near the barometer, and always read the barometer at the same room temperature. If the most common temperature in the room is 23°C, be sure that the temperature of the room is 23°C every time you record the position of the barometer's pointer. With this in mind, you should be able to trust the readings on your homemade barometer.

Resource

Tolman, M. N. 1995. *Hands-on Earth science activities* for grades K–8. West Nyack, NY: Parker.

Making a Wind Vane

By Marvin N. Tolman and Garry R. Hardy

Students can make a simple wind vane using paper and a straw and mounting it on a lead pencil to determine direction of air movement. Students will need the following materials:

- 5 cm x 15 cm piece of construction paper
- Straw
- Cellophane tape
- Hat pin
- Pencil with an eraser
- Fan
- Protective goggles
- Small bead (optional)

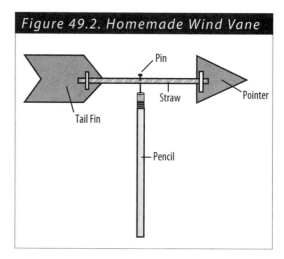

Figure 49.2. Homemade Wind Vane

Have students draw a pointer and a tail fan for their wind vane on the construction paper and cut them out. The exact dimensions are not critical, but the tail fin must be larger than the pointer (Figure 49.2). Wearing protective goggles, students need to cut a 1.5 cm slit in each end of the straw, sliding the pointer in one end and the tail fin in the other end. Have students tape the pieces in place to secure them. They now have an arrow for their wind vane.

A student can find the arrow's center of gravity (i.e., balance point) by laying the arrow on a finger and moving it back and forth until it balances. Have students mark the balance point on the straw and push the pin through the straw at that point. The pin, the pointer, and the tail fin need to line up together. Next, have students push the pin into the eraser of the pencil. If the arrow doesn't turn freely on the pin, the problem can be solved by having students turn the straw around on the pin. The hole becomes slightly larger, and the arrow should turn. It sometimes also helps to put a small bead on the pin between the straw and the pencil eraser to act as a spacer. The wind vane is now complete!

To test the wind vanes, students can hold the wind vanes in front of a fan, using the pencil as a handle. The arrow of the wind vane will point into the wind, showing which direction the wind is coming from. When the fan is turned on, the arrow should point toward the fan. The tail fin has to be larger than the pointer to offer more resistance to the wind and to force the pointer into the wind.

Take the wind vanes outdoors, and let them show which direction the wind is coming from.

Resource
Tolman, M. N. 1995. Hands-on Earth science activities for Grades K–8. West Nyack, NY: Parker.

Weather Watchers

Cloud Observation Form

Name: _____

Date: _____

When I look up, I see a cloud that looks like ...

Date: _____

When I look up, I see a cloud that looks like ...

Date: _____

When I look up, I see a cloud that looks like ...

Date: _____

When I look up, I see a cloud that looks like ...

Weather Reporting Form

Today's date

_____

Today's weather

_____

Cloud type

_____

Percent of cloud coverage

_____

Temperature

_____

Barometric pressure

_____

Precipitation

_____

Wind direction

_____

Wind speed

_____

Chapter 50

Sunrise, Sunset

By Karen Ansberry and Emily Morgan

The next time you watch the Sun rise, take a minute to think about what's really going on. You are standing on a giant ball of rock that is hurtling through space, and the spot where you are standing is rotating in the direction of a star 93 million miles away! It makes a beautiful sunrise seem even more amazing. In this trade book–inspired K–2 lesson, students observe the pattern the Sun follows as it appears to move across the sky, and in the 3–6 lesson, students model day and night and explore the need for different time zones on Earth.

Trade Books

Day and Night

By Margaret Hall, illustrated by Jo Miller
Capstone Press, 2006
ISBN 978-0-7368-6338-4
Grades K–4

SYNOPSIS

From Capstone Press's Patterns in Nature Series, this nonfiction picture book uses simple text, diagrams, and photographs to explain that Earth's rotation causes day and night. The book also depicts things that happen in the day and in the night.

Somewhere in the World Right Now

By Stacy Schuett
Dragonfly Books, 1997
ISBN 978-0-679-88549-8
Grades K–4

SYNOPSIS

This beautifully illustrated book depicts many events that could be happening in the world at any given moment. Schuett's illustrations, each overlaid on a map, include hidden details that provide hints as to where the events are taking place. A page in the front of the book explains that because of Earth's rotation, different time zones have been established throughout the world.

Curricular Connections

According to the *National Science Education Standards*, in early elementary school, students should be asked to make regular observations of day and night (NRC 1996). It is important to note that understanding the day-night cycle may be challenging for younger learners. Not only do they need to recognize that Earth is a sphere, they must also understand that the rotation of this sphere causes the day-night cycle. These ideas may be too abstract for many young children to grasp. Therefore, it is important to keep the focus of the K–2 lesson on safe *observations* of the position of the Sun in the sky. However, we include an activity in the Elaborate phase of the lesson where students make a very simple model of day and night with a globe and lamp. We view this activity as an extension and would not hold all students accountable for being able to explain how Earth's rotation causes day and night. According to the Standards, students in upper elementary and middle school should understand that most objects in the solar system are in regular and predictable motion and that these motions explain such phenomena as the day-night cycle, seasons, phases of the Moon, and eclipses. Observations of the Sun's apparent movements will provide the motivation and the basis from which students can construct a model explaining the reason for day and night. In the 3–6 lesson, we give more focus to the globe and lamp model in explaining the day-night cycle.

Grades K–2: Where Is the Sun?

Materials

- Let's Learn About Day and Night anticipation guide (p. 301)
- Chart paper
- Globe (one per group of three or four students)
- Lamp with shade removed

Engage

Give each student a copy of the Day and Night Anticipation Guide, which asks the following questions: Does the Sun appear in the same place in the sky all day long? Does the Earth spin? Does the Sun ever stop shining? Can it be day in one place on Earth and night somewhere else? Does the Sun shine highest in the sky at noon? Is the Earth shaped like a ball? Have students record their answers in the "Before" column on the left-hand side of the page. For younger students, this can be done with the whole class, with the answers recorded on the board or chart paper.

Explore

Choose an outdoor location where students can observe the Sun in the morning, noon, and afternoon. Remind students to never look directly at the Sun. Looking at the Sun can damage your eyes! Each time you take the class outside, have students face the southern sky and observe, with caution, the position of the Sun relative to a landmark (e.g., a flagpole) at each of these times for three days. Record the observations on a class chart using drawings of the landmark and the Sun.

Explain

After the third day of observations, use the evidence from the class chart to discuss the following questions: "Where did you see the Sun in the sky each morning? Noon? Afternoon? Did you notice any patterns?" (Students should notice that the Sun is always lowest in the eastern sky in the morning, highest in the sky at noon, and lowest in the western sky in the afternoon.) "At what time of day did the Sun seem highest in the sky?" (noon). "Is there any time that you go outside and can't see the Sun?" (nighttime). "Why can't you see the Sun at night? Is it still shining?" (answers will vary). Show students the cover of *Day and Night*. Tell them that this book can help them answer the question about why they can't see the Sun at night. Read the book aloud,

Sunrise, Sunset

and discuss how the Sun is always shining, but the part of the Earth that faces away from the Sun is experiencing night.

Elaborate

Introduce the author/illustrator of *Somewhere in the World Right Now,* and read the book aloud to students. Model the questioning skills of a good reader as you read by asking questions such as, "Can it be day somewhere on Earth and night somewhere else? Do you know someone who lives in a different part of the country or world where it is a different time than it is here?" Next, show students a globe and a lamp. Tell them that the globe represents the Earth. Ask, "What shape is the Earth?" Tell them the lamp represents the Sun and ask, "Does the Sun ever stop shining?" Say, "Because the Sun never stops shining, we will keep the lamp lit in the center of the room." Next, give each group of students a globe and have them use it to locate the areas of land and water, find the United States, and then pinpoint their location. Make the room as dark as possible and have each group model daytime in their location. Then ask them to name some places that are having night while they are experiencing day.

Evaluate

Revisit the Day and Night anticipation guide and have students fill in the "After" column on the right-hand side of the page. Discuss the answers and allow students to explain their thinking. Answers are (1) no, (2) yes, (3) no, (4) yes, (5) yes, and (6) yes.

Grades 3–6: Time Zones

Materials

- Clocks set at different times, representing different locations around the world
- Lamp (one per group of three to four students)
- Globe (one per group of three to four students)
- Somewhere in the World Right Now student page (p. 302)

Engage

Before class, bring in several clocks set at different times to represent actual times at different locations around the world and labeled with the city and country. You can find times of many cities around the world at *www.timeanddate.com/world-clock*. Introduce the author/illustrator of *Somewhere in the World Right Now*. Skipping "A Note to the Reader" in the front of the book (this section will be used later to provide the scientific explanation for the students), read the book aloud to students. Model the questioning skills of a good reader as you read by asking questions such as the following: "Is it true that somewhere in the world it is already tomorrow? How can the Sun be rising and setting at the same time? How can all of these things be happening in the world right now?" After reading, ask, "Do you know someone who lives in a different part of the country or world where it is a different time than it is here? Have you ever been to a place where you had to set your watch differently?" Have students examine the clocks set for different times for different places in the world. Determine students' prior knowledge and misconceptions about Earth/Sun relationships by asking them to share ideas about how it can be so many different times at the same moment. One common misconception some students may have is that the Sun actually moves across the sky, orbiting the Earth.

Explore

Provide each group of students with a lamp and a globe. Tell them that they are going to use the lamp as a model of the Sun and the globe as a model of the Earth.

Before they begin the activity, ask students the following: "How does the Earth move?" (It rotates on its axis and revolves around the Sun.) "What do the movements of the Earth have to do with how we keep time?" (One rotation is one day, and one revolution is one year.) "Which movement do you think causes day and night?" (Earth's rotation). Then give students a few minutes to explore the following question with

Sunrise, Sunset

the model: "How can it be different times in different places on the Earth?" After students have had time to explore the model, pass out the Somewhere in the World Right Now student page. Tell students to use the lamp and globe to answer the questions on the student page.

Explain

Discuss the student responses on the Somewhere in the World Right Now student page. Have students share any observations, answers, and questions they still have. Tell students you will be reading an informational page titled "A Note to the Reader" in the front of *Somewhere in the World Right Now.* Have students listen for answers to any questions they might still have about time zones, the date line, or Earth's rotation.

Elaborate/Evaluate

Tell students they will be writing and illustrating a children's picture book that can be used to explain what causes day and night and how it can be day in one part of the world and night somewhere else. Their finished products should include simple text, colorful illustrations, and clearly labeled diagrams (see p. 303). Have available some picture books about astronomy written for young children, such as *The Sun Is My Favorite Star* by Frank Asch (2008) and *The Moon Book* by Gail Gibbons (1998). Share some examples of simple text, colorful illustrations, and clearly labeled diagrams.

References

Asch, F. 2008. *The Sun is my favorite star.* New York: Harcourt.

Gibbons, G. 1998. *The Moon book.* New York: Holiday House.

National Research Council (NRC). 1996. *National science education standards.* Washington, DC: National Academies Press.

Internet Resource

The World Clock—Time Zones
www.timeanddate.com/worldclock

Sunrise, Sunset

Name: _____

Let's Learn About Day and Night

Anticipation Guide

Before
Yes or No

After
Yes or No

_____ 1. Does the sun appear in the same place in the sky all day long? _____

_____ 2. Does the Earth spin? _____

_____ 3. Does the Sun ever stop shining? _____

_____ 4. Can it be day in one place on Earth and night somewhere else? _____

_____ 5. Does the Sun shine highest in in the sky at noon? _____

_____ 6. Is the Earth shaped like a ball? _____

Name: _____   *Teacher* _____

Somewhere in the World Right Now

Place your lamp and globe about 50 cm apart with the lamp shining toward the globe.

1. What does the lamp represent in this model? _____

2. What does the globe represent? _____

3. Did you find the arrow near the equator that shows the direction Earth turns? _____ Be sure to turn your globe slowly in that direction.

4. Model daytime in your state.

5. List two places that are having night when it is daytime in your location. _____

   _____

6. Model sunrise in your state by turning the globe so that your state is just entering the lamp's light.

7. Can the Sun be rising and setting at the same time? Explain.

   _____

   _____

   _____

   _____

☐ *Teacher Checkpoint*

Name: _____

Make a Picture Book

Write and illustrate a children's picture book that can be used to explain what causes day and night and to explain what causes the Sun to *appear* to move across the sky each day.

Books should include:

1. a catchy title to grab the attention of the reader;

2. an accurate explanation and clearly labeled diagram showing what causes day and night;

3. an accurate explanation and clearly labeled diagram of what causes the Sun to appear to move across the sky each day;

4. simple text a young child could understand; and

5. colorful illustrations.

Be creative! Have fun!

Chapter 51

Moon Phases and Models

By Karen Ansberry and Emily Morgan

From the time they are very young, children are naturally curious about the Moon. They may wonder about the different shapes of the Moon when they look up at the night sky. In this primary lesson, students discover through direct observations and reading that the Moon's shape follows a pattern. In the upper-elementary lesson, students explore the reason for this pattern using a model.

Trade Books

Phases of the Moon
By Gillia M. Olson, illustrated by Jo Miller
Capstone Press, 2008
ISBN 978-0-7368-9617-7
Grades K–4

SYNOPSIS
Simple text and photographs introduce Moon phases, including why they occur and what they are called.

The Moon Book
By Gail Gibbons
Holiday House, 1998
ISBN 978-0-8234-1364-5
Grades K–4

SYNOPSIS
The Moon Book *identifies the Moon as our only natural satellite, describes its movement and phases, and discusses how we have observed and explored it over the years.*

Curricular Connections

According to the *National Science Education Standards*, the focus of space science in the early years should be on the idea that objects in the sky have observable patterns of movement. By observing the day and night sky regularly, early elementary students can learn to identify sequences of changes and look for patterns in these changes. Specifically, when it comes to the Moon's phases, students should notice that "the observable shape of the Moon changes from day to day in a cycle that lasts about a month" (NRC 1996, p. 134). The lesson for grades K–3 students is limited to making observations, developing descriptions, and finding patterns; whereas grades 4–6 students use a model to explore the concept that Moon phases are caused by the Moon's orbit around Earth.

Grades K–3: Moon Monitors

Materials

- Large O-W-L chart
- Sticky notes
- My Moon Journal (p. 309)

Engage

Show students the photographs of the Moon from *Phases of the Moon* without reading the text. Ask them what they notice or wonder about the Moon. Prepare a large O-W-L (Observations, Wonderings, Learnings) chart and keep it posted prominently in the classroom for the duration of the unit. First model your own wonderings about the Moon by writing the following questions on the O-W-L chart under the W (wonderings) section: Does the Moon really change shape? Where does the Moon's light come from? Does the Moon's shape follow a pattern? How long does it take for the Moon to go through its changes? Then, invite students to write their own questions about the Moon on large sticky notes to post in the wonderings section of the O-W-L chart.

Explore

Ask students how they might find the answers to some of their questions. Discuss that scientists find answers by making careful observations, doing experiments over and over, communicating with other scientists, and so on. Tell students that they can find out more about the Moon by observing it every evening for a month. Give each student a copy of My Moon Journal. Ask them to look at the Moon each night and draw what it looks like (if it can be seen). In the classroom, keep a daily bulletin board of the Moon phases for a month. Ideally, students should make their own observations of the Moon for at least a month, but Moon calendars can also be downloaded at *www.stardate.org*.

- **Note:** A common misconception about the Moon is that the Moon gets larger and smaller. Empty circles on the Moon Journal student page are provided so that students can darken the areas of the Moon that are not lighted. This method of recording Moon phases takes into account that the entire Moon is present, even if some of its surface cannot be seen.

Explain

Discuss students' observations throughout the month using some of the following questions: "Was the Moon the same shape each time you saw it?" (no). "Did you see the Moon every time you looked for it?" (no). "Was the Moon in the same place in the sky each time you saw it?" (no). "On a cloudy night, how can you tell if the Moon is still there?" (You can see moonlight behind the clouds.) "Can we ever see the Moon in the daytime?" (yes). "What did the Moon look like on the first night of your journal? On the last night?" (Students should notice that the Moon's shape is the same on the first night

Moon Phases and Models

and the last night of the journal.) "When you look at your journal, do you see any patterns?" (Students may notice a pattern of the Moon changing shape.) During this discussion, students should be able to explain their observations and compare them to the observations of others.

Elaborate

Revisit *Phases of the Moon*, and explain that this nonfiction book might help them make more sense of their Moon observations. Ask students to listen for any answers to the "wonderings" on the O-W-L chart as you read the book aloud. After reading, write the answers in the L column of the O-W-L chart and discuss.

Evaluate

Ask, "How does the pattern of the phases you observed in your Moon journal compare to the pattern of the phases in the book?" (The phases recorded over one month in our journals are in the same order as the phases we saw in the book.) Students should notice that the phases of the Moon occur in a certain order and have names. Ask students to find one example of each phase on their journal and write its name beneath the picture (full, gibbous, quarter, crescent, new).

Grades 4–6: Moon Modeling

Materials
- Moon Survey student page (p. 310)
- Lamp with shade removed
- Smooth foam spheres (one per student)
- The Changing Moon student page (p. 311)

Engage

Have students complete The Moon Survey student page and discuss the results of their surveys before they begin the next activity. Ask the following questions as you discuss the surveys: "What are some of the answers you received? Are there any answers

that you think are wrong? Why? What do you think is the correct answer to the question on the survey?"

Explore

Now that students have heard a lot of different ideas people have about why the Moon looks different from night to night, tell them that they can explore this question using a model. You'll need a lamp and one foam ball for each student.

Darken the room—the darker, the better. Give each student a foam ball stuck on the end of a pencil. Explain that the foam ball is a model of the Moon, the lamp is a model of the Sun, and their heads represent Earth. Before the guided activity below, give students time to explore the model and test different ideas about what causes Moon phases. Next, guide students through the following activity to model how the Moon changes shape:

- With their faces toward the lamp, students hold the balls slightly above their heads so that they have to look up a little to see them. In this position, students cannot see the lighted side of the ball. Tell them that this is called a new Moon.

- Tell students to move their arms holding the balls slightly to the left while still looking at the balls and holding them a little above their heads. Tell them that when we see this sliver of the lighted side of the Moon, we call it a crescent Moon.

Ask, "Where does the Moon's light come from?" (The light is coming from the Sun and is reflected off the Moon.) "Some people think the Moon phases are caused by the Earth's shadow. How does this model provide evidence that this idea is not correct or supported?" (The shadow of my head, which represents the Earth, is not on the Moon in this position. It is behind me.)

Instruct the students to keep turning to the left and soon they will see more of the lighted half of the balls. Tell them that this is called a quarter Moon.

- Have them turn a little more and almost all of the balls will be lit. Tell them that this is called a gibbous Moon.

- Students can keep turning until they see all the lighted half of the balls. Tell them that this is called a full Moon.

- As students continue to turn in the same direction, they will see less and less of the lighted part of the ball. First they will see a gibbous Moon, then a quarter Moon, then a thin crescent Moon, and finally they will be back to the new Moon.

- Tell students that this orbit the Moon takes around the Earth is completed in about a month.

Explain

Have students work with partners to repeat the Moon modeling activity and explain to each other the reason we see each phase of the Moon from Earth. Tell students that scientists often use their observations in combination with models to develop explanations of scientific events. Ask, "What explanations can we develop from our month of Moon observations and the Moon modeling activity we just did?" (The Moon phases occur in a regular pattern. The orbit of the Moon around the Earth causes the phases.)

Elaborate

Tell students that in science it is important to check what you have learned through observations and models with what is known in the scientific world. Ask students to compare their ideas and explanations about the Moon phases to the information presented in *The Moon Book*. Read the book aloud, stopping every so often to allow for discussion as you read. After reading the section on eclipses, challenge students to use the foam ball and lamp model to demonstrate lunar and solar eclipses. Then ask, "What Moon phase does it need to be in order for a lunar eclipse to occur?" (full Moon). "What Moon phase does it need to be in order for a solar eclipse to occur?" (new Moon).

Evaluate

Give students copies of The Changing Moon student page (p. 311), where they will draw and label two diagrams: Earth, Moon, and Sun with the Moon in the new Moon phase and another with the Moon in the full Moon phase. Using the evidence from the moon modeling activity and their diagrams, they will answer the question, "Why does the Moon seem to change shape?"

Reference

National Research Council (NRC). 1996. *National science education standards.* Washington, DC: National Academies Press.

Moon Phases and Models

My Moon Journal

Dates of Observation

Sunday	Monday	Tuesday	Wednesday	Thursday	Friday	Saturday
◯	◯	◯	◯	◯	◯	◯
◯	◯	◯	◯	◯	◯	◯
◯	◯	◯	◯	◯	◯	◯
◯	◯	◯	◯	◯	◯	◯

Name:_____

Moon Survey

Ask three people the following question and record their answers on the lines below:

What causes the Moon to look different throughout a month?

Person 1

_____

_____

_____

_____

Person 2

_____

_____

_____

_____

Person 3

_____

_____

_____

_____

Moon Phases and Models

Name: _____

The Changing Moon

Draw and label a diagram of the Earth, Moon, and Sun with the Moon in the New Moon Phase.

New Moon

Draw and label a diagram of the Earth, Moon, and Sun with the Moon in the Full Moon Phase.

Full Moon

Why does the Moon seem to change shape? Use evidence from the Moon modeling activity and your diagrams above to answer.

_____
_____
_____
_____

Chapter 52

Seeing Stars

By Christine Anne Royce

The winter months are a great time to make observations of several familiar constellations. While there's no scientific reason to know the constellations—they are simply imaginative pictures imposed on stars—studying constellations can help students connect with culture in a fun way and develop the awareness that stars are different in apparent brightness and color. Exploring the night sky over a period of weeks can also help students notice the motion of the Sun, Moon, and planets.

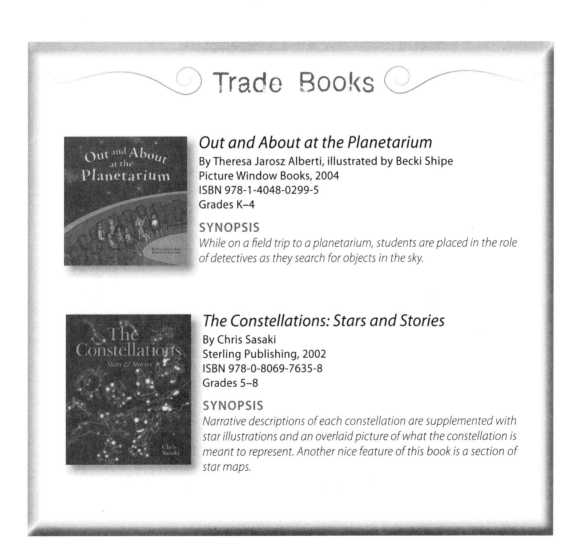

Trade Books

Out and About at the Planetarium
By Theresa Jarosz Alberti, illustrated by Becki Shipe
Picture Window Books, 2004
ISBN 978-1-4048-0299-5
Grades K–4

SYNOPSIS
While on a field trip to a planetarium, students are placed in the role of detectives as they search for objects in the sky.

The Constellations: Stars and Stories
By Chris Sasaki
Sterling Publishing, 2002
ISBN 978-0-8069-7635-8
Grades 5–8

SYNOPSIS
Narrative descriptions of each constellation are supplemented with star illustrations and an overlaid picture of what the constellation is meant to represent. Another nice feature of this book is a section of star maps.

Curricular Connections

"Twinkle, twinkle, little star, How I wonder what you are!" This wonder itself demonstrates children's "powers of observation" as they look up into the night sky. Having a well-developed process skill of observation is important when dealing with astronomy concepts at the elementary level. The Standards state that the appropriate concept for children at the younger grades is that "Objects in the sky have patterns of movement" (NRC 1996, p. 134), forming a basis for later understanding of the relationships among the Earth, Sun, Moon, and the rest of the solar system. Many classroom activities at this level encourage students to observe objects and materials in their natural environments to build the skill of observation; students are naturally interested in "looking up" at what is found in space. The *National Science Education Standards* recommendation to observe the sky and observe changes lends itself nicely to observing patterns associated with constellations.

The first activity places children in a comfortable environment while being guided through making "observations" of constellations. As children get older and become more familiar with the idea of patterns, change, and making observations, they can be provided with not only in-class opportunities to discuss constellations but also the opportunity for take-home assignments related to stargazing.

Grades K–3: Creating Constellations

Purpose

Using images of constellations, the internet, and a planetarium (if possible), students will make observations about patterns that the stars form.

Materials

- Pictures of constellations
- Black paper
- White chalk

Procedure

1. Begin by asking students to listen to *Out and About at the Planetarium*. They should place themselves in the role of "space detectives" as the story suggests. As space detectives, they will be looking for or making observations about the stars in the night sky.

2. The first part of the book focuses on planets, the Sun, the Moon, and the parts of a planetarium. (If your school district or local community has a planetarium, this would be a great time to schedule a visit.) Beginning on page 14 of the book, the story focuses on stars—how some are brighter than others and some form "pictures" in the sky.

3. Show the students pictures of various constellations via a projector or printouts.

4. As you read the book, at first point out the different stars for each picture within the overlaid image. Ask the students to observe what part of the picture each star might represent—for example in Orion the Hunter, the three stars in a row represent the belt and the other main stars represent the shoulders and knees. Some constellations will be easier to see than others. Ask, "Have you ever observed the night sky or a constellation? If so, which one?" and "Do you think it was easy for ancient people to observe the night sky?"

5. This is a great point to introduce how the constellations obtained their names and perhaps some of the stories behind the constellations, which connects the history of science to this lesson.

6. After students have had a chance to observe the constellations either in a classroom or in a planetarium, ask the students to generate a list of things they learned. Some items may include, "Constellations are pictures in the sky," "Different stars make up different constellations," or

"Not all constellations are easy to recognize." The key point to this activity is to have students make observations about what they see and connect those observations to the larger picture.

7. After students had a chance to observe the constellations, ask them to create their own constellations using a piece of black paper and white chalk. The students should think about what they know about constellations—that they have stars at key points in the picture; that the stars might be different brightness (which is often represented by size in an illustration); and that there is usually a story that goes along with the constellation.

8. Allow each student to design his or her own constellation by drawing it on the paper and lightly drawing the overlaid picture of what it might represent. Students can then write out the story that accompanies their constellations or narrate the story using a recorder. Constellations can be displayed for your own classroom collection.

Grades 4–6: Stargazing at Home

Purpose

Older students and their parents have an opportunity to engage in some actual stargazing using starfinder maps.

Materials

- Starfinders that the students will assemble (See Internet Resources)
- Seeing Stars student page (p. 317)

Procedure

1. *The Constellations: Stars and Stories* is a great book to use as a reference for students. Teachers should select several constellations that they want the students to focus on for this activity.

The selection of constellations depends on the location on the Earth, as well as the time of the year. For example, for the northern hemisphere in February, the following constellations would be good choices: Gemini, Orion, Canis Major, Canis Minor, and Cassiopeia.

Teachers should engage the students in a conversation about what a constellation is, how to locate them using a star finder (described in the book), and the fact that different constellations are visible at different points during the year. This can be done while creating the star finder wheels and modeling their use.

A letter should be sent home explaining that this project will need parental involvement. Students should be reminded never to venture into dark areas without a parent and to wear reflective clothing. Students should use the Seeing Stars student page to sketch the constellations they see. Students can record their observations about the brightness of the stars in each constellation, as well as the position of the constellations at different points in the evening. Students should be encouraged to initially go out at a certain time to allow them to all have similar observations as to the position of the constellation.

Teachers may want to check with local astronomy groups or clubs to see if they are willing to set up "star parties" for school groups.

2. Explain to the students that they are going to be making observations of different constellations over the next several weeks and ask them to make some predictions about other things they may observe in the night sky. Some possibilities include clouds, airplanes, meteor showers, or planets. Each clear night, the student should look at one or more constellations and sketch what they see on the Seeing Stars student page (p. 317). Change the time they observe the constellations by 15 minutes each night, thus allowing some change in position due to the rotation of the Earth.

3. Students should bring in a drawing of the previous night's sky each day. After students have had an opportunity to compare the pictures, ask them what they noticed about the position of the constellations each night. Other questions that could be discussed at this point include, "Do you think there are different constellations visible at different points in the year?" or "Why do you think there are different seasons' constellations visible in the northern and southern hemispheres?"

Teachers can have students practice finding constellations against a star field by allowing students to use a game that challenges students to find different constellations (see Internet Resources).

Reference

National Research Council (NRC). 1996. *National science education standards.* Washington, DC: National Academies Press.

Internet Resources

Constellation Hunt
www.kidsastronomy.com/astroskymap/ constellation_hunt.htm,

Interactive Sky Map
www.kidsastronomy.com/astroskymap/ constellations.htm,

Make a Star Finder
http://spaceplace.nasa.gov/en/kids/ st6starfinder/st6starfinder.shtml

Seeing Stars

Name: _____

The constellation you picked to observe is _____ .

Sketch the constellation for each day you observe it. Make the dots bigger for brighter stars and smaller for dimmer stars.

First date: _____ *Time:* _____	*Second date:* _____ *Time:* _____

Third date: _____ *Time:* _____	*Fourth date:* _____ *Time:* _____

Did you notice any difference over the days you observed the constellation? _____

_____

_____

Other objects you observed in the night sky were _____

_____

_____

Index